Praise for *The Vertical Farm*

"A book you will read, and then you will read it again. It is a book that could begin a revolution. Let's hope it does. . . . *The Vertical Farm* is easily accessible to subject novice and expert alike. . . . Not at all a how-to book for small-scale change, the plan described will influence the future of more than just farming. Despommier invites the reader to realize a bigger future of sustainability, even better than we can imagine. . . . If you are anyone who is looking for a more hopeful future, you need to read this book."

—*Chicago Examiner*

"Dickson Despommier is a futurist, an architect, and an intellectual in the same vein as Leonardo da Vinci, I. M. Pei, and Buckminster Fuller. Vertical farms will be remembered as one of the preeminent breakthroughs of the early twenty-first century, and Despommier will be remembered as the man who brought them to us."

—Josh Tickell, director of *Fields of Fuel*,
winner of the 2008 Sundance Film Festival
Audience Award for Best Documentary

"Cities of the future must generate their own food supply. Dickson Despommier's elegant, simple answer for achieving this goal is vertical farming. Welcome to the third green revolution."

—Peter Diamandis, chairman of the XPRIZE Foundation
and cofounder of Singularity University

"Despommier looked well outside the box when he began investigating ways to improve the growing and distribution of food crops. The result is a revolutionary theory that could allow the growth of food crops 24 hours a day, 365 days a year; protect crops from weather; reuse water; provide jobs for local people; eliminate the need for pesticides, fertilizers, and herbicides; reduce dependence on fossil fuels; prevent crop loss due to shipping and storage; and do away with agricultural runoff. The answer is the vertical farm."
—*The Globe and Mail* (Toronto)

"Persuasive . . . Given Dr. Despommier's scientific background we might expect this book to be a dry recitation of facts and figures, but nothing could be further from the case. Despommier writes passionately and argues, at times, even stridently."
—*The New York Journal of Books*

"A visionary known the world over, Despommier believes that the 'vertical farm is the keystone enterprise for establishing an urban-based ecosystem' and for 'restoring balance between our lives and the rest of nature.' . . . Provocative."
—*Booklist* (starred review)

"A captivating argument that will intrigue general readers and give policy makers and investors much to ponder."
—*Kirkus Reviews*

DR. DICKSON DESPOMMIER

THE VERTICAL FARM

DR. DICKSON DESPOMMIER spent thirty-eight years as a professor of microbiology and public health in environmental health sciences at Columbia University, where he won the Best Teacher Award six times. In 2003, he received the American Medical Student Association's National Golden Apple Award for Teaching Excellence. His work on vertical farms has been featured on such top national media as the BBC, French national television, CNN, *The Colbert Report*, and *The Tonight Show*, as well as in full-length articles in *The New York Times*, *Time* magazine, *Scientific American*, and *The Washington Post*. He has spoken at the TED Conference, PopTech, and the World Science Festival, and has been invited by the governments of China, India, Mexico, Jordan, Brazil, Canada, and South Korea to work on their environmental problems. He has been invited to speak at numerous national and international professional annual meetings as a keynote speaker, and at universities, including Harvard and MIT. He is one of the visionaries featured at the Museum of Science and Industry, Chicago. Dr. Despommier lives in Fort Lee, New Jersey.

THE VERTICAL FARM

THE VERTICAL FARM

FEEDING THE WORLD IN THE 21ST CENTURY

DR.
DICKSON
DESPOMMIER

PICADOR
A THOMAS DUNNE BOOK
St. Martin's Press
New York

Picador
120 Broadway, New York 10271

Printed in the United States of America
Published in the United States 2010 by Thomas Dunne Books, New York
First Picador paperback edition, 2011
Second Picador paperback edition, 2020

The text stock contains 20 percent post-consumer waste recycled fiber.

Library of Congress Control Number: 2010029257
Second Picador Paperback ISBN: 978-1-250-76980-0

Designed by Gretchen Achilles

1 3 5 7 9 10 8 6 4 2

For the more than one billion people who, through no fault of their own, go to sleep hungry each night; and for the three billion more who will most likely arrive on this planet over the next forty years to join them in their suffering, if nothing changes.

CONTENTS

FOREWORD

America has had a love-hate relationship with "the farm" going back for many decades. It now looks as though our relationship to farming has a chance to become more mature, both technologically and culturally, in the twenty-first century.

The farm can be that romantic, Rockwellian vision of a simple life, somehow more authentic than the urban/suburban life—which is how most of us live now. We like to see a logo with that caricature image of a little red barn with a silo and paddock next to it on the packaging of our dairy products, eggs, cereals, berries, mushrooms, bacon, etc. We like to think of the "farmer" as someone possessing good old-fashioned American values and common sense, with a big family of healthy wholesome children who all go to church every Sunday. An image of people like that, producing our food somewhere, makes us feel better than if we saw where most of our food was actually coming from.

But the farm is equally a source of derision and mockery. For both recent immigrants and multigenerational Americans, "country" can mean unsophisticated. The people we trust to produce enough safe, clean food for us to eat are given little respect on the whole. For many decades, the smart kids were sent off to the city; the less smart stayed back on the farm. It was normal for an individual to move from the farm to the city, but nobody grew up in the city to become a farmer.

Some Americans associate working the land with slavery. A great tide of African Americans migrated from the South to the industrial cities of the North to escape the land and everything associated with it. They found better-paying jobs in factories, and those twentieth-century manufacturing jobs gave Blacks a chance to realize middle-class stability and ambitions. The land was something you left behind. In the minds of many of our relatives and ancestors, the more distant you were from that past, the better.

Much of my work over the past decade has centered on horticultural infrastructure with regard to managing stormwater runoff and urban heat-island effect. Things like green roofing, urban forestry, and wetland and estuary restoration are all important parts of the solution to those environmental challenges—and they provide great jobs for people who often have difficulty getting a job. But many times when I deliver this message to inner-city communities that are suffering from severe unemployment as well as the health effects of environmental mismanagement, there is resistance. Working the land is seen as a step back, not a step forward.

Now that the majority of manufacturing jobs have moved overseas, and our agriculture system has long since abandoned the bucolic notion of a "family farm" in favor of spreadsheet economics, we have a remarkable chance to revise many of those old assumptions, and reexamine the good and bad legacies of our agricultural past. We can look at our needs and the technology available to meet those needs in ways that treat the land and the people with more respect and dignity.

To be fair, our current food production and distribution system does deliver edible calories to the people at an affordable "price." But its toll on both the environment and its consumers is astonishing. The herbicides and pesticides that are applied to the plants wash out into our rivers and oceans—creating dead zones where fishing is no longer viable. That means hard-working people might not have jobs in the seafood industry because of agribusiness decisions made way upstream, decisions that are often subsidized by our tax dollars.

Chemical fertilizers leave the soil incapable of supporting plants, without even more fertilizer. All these chemicals can eventually make it to the drinking water as well. For folks living near the trans-shipping centers where the food comes into a city and goes out to grocers, restaurants, etc., the diesel exhaust is thick in the local air, and the lungs of anyone who cares to breathe. Children living near these facilities are put at risk by the large trucks passing through their neighborhoods, discouraging active play and exacerbating the obesity epidemic facing our nation. These quality-of-life costs are borne most often by poor people, with no just compensation.

In the time between now and the realization of Dickson Despommier's vision for our food system, there are many opportunities for innovation and entrepreneurship. If the skyscraper farm is like a 747 jetliner, we are now at the stage of the Wright Brothers. All kinds of urban micro-agribusinesses are bursting onto the scene in cities across America, and in other countries around the world. There will be many failures as a legion of tinkerers and engineers all struggle to take off with the right combination of profitability, sustainability, and quality food. Exciting career ladders are a part of the mix—this work will be perfected over time with more and more minds engaged. New distribution and delivery systems, and on-the-vine inventory practices will create competitive arenas with relatively low barriers to entry due to the shorter geographic distances between producer and consumer.

Those jobs will bring hope to some of the Americans who once would have been employed by jobs that have been exported. By dramatically reducing the financial burden of fossil fuel–based transport, refrigeration, and chemical inputs, more money can stay in local urban economies where food is produced—and more money will be available for jobs in our cities.

The first of these farms will likely be located where the land is least expensive, and that usually means poor neighborhoods. It will be many years before urban farming production capacity even approaches filling the demand out there for food—which means a steady increase for this sector is possible for many years to come. That means many years of living, local, positive examples for economic prosperity that can inspire and employ people for generations. This productive commercial activity will be a welcome relief from the type of economic development we generally see driven into low-income neighborhoods—low wage retail, waste handling facilities, stadiums, and jails. It's time we stopped building such tributes to our collective failure to innovate, and embrace our fellow Americans. It's time to build monuments to hope and prosperity. Vertical farming represents an elegant opportunity for us to rise to that challenge.

Majora Carter

Majora Carter founded Sustainable South Bronx in 2001 to achieve environmental justice through economically sustainable projects informed by community needs. Her work has earned numerous awards including a MacArthur "Genius" Fellowship; she was also named one of Essence Magazine's Twenty-five Most Influential African Americans. She is a board member of the Wilderness Society and hosts a special Corporation for Public Broadcasting radio series "The Promised Land" (thepromisedland.org). She is currently President of the Majora Carter Group, LLC, a green-project development and consulting group.

THE VERTICAL FARM

INTRODUCTION

Fifteen thousand years ago, there was not a single farm on the planet. Fast-forward to the present, when we now farm a landmass the size of South America, which does not include grazing land. Along the way, we invented, among other things, written language, mathematics, music, and, of course, cities. Yet our journey from hunter-gatherers to urban dwellers still hasn't produced a single metropolis that is truly healthy to live in.

As populations grew and urban life became the norm, our habit for producing mountains of waste began to take its toll. Garbage provided sustenance for a wide variety of peri-domestic diseases that emerged and then became endemic. For example, in the twelfth century, trash of all kinds, strewn carelessly across the European landscape by returning crusaders from the Middle East, attracted hordes of rats. These vermin harbored the plague bacillus, a flea-borne infection. As the rats died, their fleas soon found human hosts to feed on, igniting the first outbreak of the Black Death in Europe. It killed more than one-third of all those living there. Cholera came to Europe in 1836 by way of trading vessels from the Bay of Bengal, first to London, England. Because of the high nutrient content of the Thames River, due mostly to garbage dumping, cholera became endemic, killing thousands of Londoners every year until John Snow figured out its modus operandi.

You'd think we would have learned something from all this. But as late as the nineteenth century, waste on the streets of New York City was still causing massive outbreaks of diarrheal diseases. To this day, most cities still haven't found a good use for garbage. New York remains plagued by vermin and poor-sanitation-related diseases such as asthma. With landfills for most cities now bulging at the seams, urban communities will have to reinvent waste management. Yet there is hope. All of this is about to change. We now have in our hands the tools and the desire to convert squalid urban blight into places where we'd want to raise our children. Once we have transformed our urban centers, we can turn our attention to renewing the hardwood forests that we destroyed in our zeal to create the farmlands that now produce food for our cities.

Sustainable urban life is technologically achievable, and most important, highly desirable. For example, food waste can easily be converted back into energy employing clean state-of-the-art incineration technologies, and wastewater can be converted back into drinking water. For the first time in history, an entire city can choose to become the functional urban equivalent of a natural ecosystem. We could even generate energy from incinerating human feces if we so desired. We have the ability to create a "cradle to cradle" waste-free economy. All that is needed is the political will to do so. Once we begin the process, cities will be able to live within their means without further damaging the environment.

VERTICAL FARMS

Repairing the environment and still having enough good, healthy food choices may seem like mutually exclusive goals. If the world's population continues to increase, wouldn't we need to cut down even more forest to produce enough food to feed everyone? Not necessarily. One solution lies in vertical farms. These farms would raise food without soil in specially constructed buildings. When farms are successfully moved to cities, we can convert significant amounts of farmland back into whatever ecosystem was there originally, simply by leaving it alone.

This plan may sound naive and impractical. Yet the concept of vertical farming is dead simple. Still, making it happen could require the kind of technical expertise needed for, say, rocket science or brain surgery. Then again, human beings do rocket science and brain surgery quite well. We should not shy away from the challenge of farming vertically simply because it requires cutting-edge engineering, architecture, and agronomy. All of this is within our grasp. We understand the hydroponic and aeroponic farming methodologies needed to grow crops within multistory buildings. Although there are still no examples of functioning vertical farms, many urban planners have become familiar with the concept and are now looking for ways to make it happen. There are already plans on the drawing board by developers in wealthy countries that are running short of arable farmland. In other places where food is becoming scarce and people are going to bed hungry, vertical farms could eventually solve this seemingly intractable problem.

The idea of growing crops in tall buildings might sound strange. But farming indoors is not a new concept. Commercially viable crops such as strawberries, tomatoes, peppers, cucumbers, herbs, and a wide variety of spices have made their way from commercial greenhouses to the world's supermarkets in ever-increasing amounts over the last fifteen years. Most of these greenhouse operations are small in comparison to the large commercial farms of the American Midwest, but unlike their outdoor counterparts, greenhouse facilities can produce crops year-round. Fish, as well as a wide variety of crustaceans and mollusks, have also been raised indoors. Chickens, ducks, and geese could conceivably be raised in indoor farms as well.

Vertical farms are immune to weather and other natural elements that can abort food production. Crops can be grown under carefully selected and well-monitored conditions that ensure optimal growth rates for each species of plant and animal year-round. In other words, there are no seasons indoors. The efficiency of each floor of a vertical farm, one acre in footprint, could be equivalent to as many as ten to twenty traditional soil-based acres, depending upon the crop. Vertical farms offer many environmental benefits as well. Farming indoors eliminates the need for fossil fuels now used for plowing, applying fertilizer, seeding, weeding, and harvesting.

CITIES WITHOUT WASTE

The ingredients in the dinner you just ate at your favorite restaurant likely came from more than fifteen hundred miles away. If you had a vertical farm in your city, all the food on your plate could come from down the block, saving huge amounts of fossil fuel now used to refrigerate and ship produce from all over the world. Also, think of what happens to the food you left on your plate. These leftovers, plus the waste generated in the food-preparation process, are currently nonrecoverable costs—also known as dinner for vermin. Now imagine if this organic waste could be converted back into energy. This would allow restaurants to be paid for the recoverable energy from their waste streams. An industry with a notoriously small (2–5 percent) profit margin would be able to earn additional income without raising the prices on its menus.

WATER: CLEAN AND CLEAR

One of the greatest urban health threats comes from liquid municipal waste (black water, which is composed, in part, of urine and feces). To disarm its potential for causing disease, it is first aerated, a process that breaks the solids into smaller and smaller particles, reduces the biomass, and converts most of the solids to oxygen-consuming bacteria. The mixture is then digested in the absence of oxygen, releasing a significant amount of methane, which some facilities are equipped to collect and use as an alternate energy source. The resulting sludge is culled and used in landfills, while the remaining grey water is chlorinated and discharged into the nearest body of water. In less developed countries, grey water is discarded without treatment. This practice greatly increases the risk of salmonella, cholera, amoebic dysentery, and other infectious diseases being transmitted by fecal contamination. In either case, it's a shameful waste of freshwater.

In many cities throughout the United States, sludge derived from wastewater treatment plants is processed further, then turned into high-grade topsoil and sold to farming communities. The cities of New York and Boston, for example, have operating sludge-to-fertilizer schemes. The problem is that most municipal sludge is often heavily contaminated by copper, mercury, zinc, arsenic, and chromium, which limits its use for farming.

Some vertical farms could act as stand-alone water-regenerating facilities. A cold-brine pipe system could be engineered to aid in the condensation and harvesting of moisture released by the plants. Plants in the vertical farm could convert safe-to-use grey water into drinking water by transpiration. The fact that the entire farm would be a closed-loop system would allow us to recover this unrealized, highly valued resource.

The resulting purified water would then be used in other vertical farms for raising fish and even algae, and for growing commercial crops. Ultimately, any water that emerged from the vertical farm would be drinkable, thus bringing it back into the community that brought it to the farm to begin with. New York City discards some one billion gallons of treated grey water every day into the Hudson River estuary. If industrial-quality water costs five cents a gallon (a conservative estimate), reclamation would be well worth the effort, even if the system cost as much as $30 billion to construct and manage. This is hardly a pie-in-the-sky scheme. A portion of Orange County, California, with a population of approximately 500,000, converts grey water back into tap water using a state-of-the-art purification system that cost its taxpayers $500 million to install. It was worth every penny.

THE END OF POLLUTION

The most pressing case for urban agriculture lies in our failure to handle waste, in particular agricultural runoff (left-over irrigation water laden with pesticides, herbicides, fertilizer, and silt). Agriculture is responsible for more ecosystem disruption than any other kind of pollution. What's more, today's farmers can't do much about it: Floods dictate the timing and extent of runoff.

Some 70 percent of all available freshwater on earth is used for irrigation, and the resulting unused portion is returned to countless rivers and streams. Runoff that reaches the oceans disconnects other ecological systems. Nitrogen fertilizer (ammonium nitrate) has the chemical property of absorbing oxygen from water. Agricultural runoff reduces the vibrant, abundant undersea life of coral reefs to barren remnants. Deforestation for purposes of freeing up farmland reinforces this toxic cycle by adding more nitrogen fertilizer to the mix and by further reducing the earth's capacity to sequester carbon from the atmosphere.

In a city with vertical farms, waste will be replaced with the recovery of unrealized energy. In nature, there is no waste. In the new eco-city, discarding anything without finding another use for it would be quite unthinkable. Imagine how absurd it would be to siphon off a gallon's worth of gasoline from the family car and pour it down the sewer. Yet this is equivalent to what we are doing with everything we now throw away.

CITIES OF THE NEAR FUTURE

Today's cities fail to meet even the minimum standards of self-reliance. No city lives within its own means. Everything consumed is produced outside the city, and as a result, waste accumulates at an alarming rate. A midsize city annually produces gigatons of solid material and billions of gallons of wastewater. Add to that the billions of dollars spent annually trying to get rid of this unwanted material, and you have a clear picture of our current environmental crisis.

But it doesn't have to be this way. Technology continues

to astound us with new proof of the inventiveness of the human species. Computers keep getting faster and more sophisticated. We are contemplating establishing colonies on the moon and on Mars. We have even collected the dust emitted from the tail of a comet. Yet despite this astounding prowess, most of the earth's inhabitants remain oblivious to the profound, and largely negative, effect they have on the planet. We continue to urbanize without building cities that are equipped to handle their populations. Most evolutionary biologists agree that continued failure to live within our means will relegate the human species to the fossil record.

Science has led the way in helping us to understand the toll we are taking on the planet. Satellites report on the status of many of the factors that contribute to climate change. Ground-based and satellite observations of coal-burning power plants, for example, support the unavoidable conclusion that we are the root cause of it. Now that we've identified the problem, we can devote our energy to finding a set of solutions. Producing food crops in mass quantities within the city limits would be a step in the right direction. The good news is that many of us are already trying to repair the environment through scientific research and philanthropic support. This is good evidence of our ability to behave in a selfless and altruistic manner when we have the opportunity to do so.

It's time to accept our connectedness to the rest of the

natural world. There is only so much natural capital out there, and we are on the verge of exhausting it. Building self-sustaining cities now will allow the land to heal itself, thereby restoring balance between our lives and the rest of nature.

CHAPTER 1

REMODELING NATURE

Nothing endures but change.

—HERACLITUS

DOWN ON THE FARM

Ten to twelve thousand years ago, all over the globe, humans began systematically to modify their environment by purposely domesticating parts of the natural world to meet their basic biological needs. Creating a reliable source of food and water was at the top of their list. Apparently, as if a switch had been thrown, we nearly unanimously tired of hunting and gathering. We learned how to grow crops derived from wild plants (corn, wheat, barley, rice) and to selectively breed various four-legged animals into tame versions of their wild counterparts for food, transportation, and, of course, labor. We catapulted out of the biosphere and into the technosphere, where we now find ourselves deeply embedded. Along the way, all natural systems suffered under our heavy foot of progress. It's the "progress" part of our history that we are currently having a problem with; the environmental crises of today have their roots deeply embedded in that last bit of human evolution. To understand the cumulative negative effects we have had on the natural world since we began to urbanize, we must first understand the essence of what the world was like without us in it (for glimpses of its former glory, see the BBC production *Planet Earth*; to see what the world might become again if we were to suddenly disappear, see Alan Weisman's book *The World Without Us*). By grasping the basics of what allows natural assemblages of plants and animals to organize into mutually dependent networks called ecosystems, we gain insight into how a city might be redesigned to mimic that process. It is my contention that if the built environment could behave by reflecting the integration of functions equivalent to that of an ecosystem, life would be a lot more bearable for all of us, and more economically stable, too.

NATURE'S MANIFESTO

The biosphere matured when terrestrial plants and animals became mutually dependent upon other each other in a harmonic symbiotic relationship. This took place over billions of years of evolutionary history. One current theory as to how all this happened, proposed first by Lynn Margulis and James Lovelock, who termed it the "Gaia hypothesis," suggests that once primitive life on earth arose, it began to modify the environment to suit its own needs. Today, most geochemists and ecologists would agree that this theory is the most reasonable explanation for how nutrients become recycled, down to how the ambient temperature of the entire planet is maintained. Symbiosis became the norm and now defines all of nature. Virtually every living thing can be shown to be dependent (either directly or indirectly) upon all other living things, except perhaps for those microbial extremophiles that live off the scant nutrients stored in solid rock. All green plants are able to grow and reproduce using only the energy contained in sunlight, together with water and a few (at least sixteen) essential minerals they obtain from the solid substrate (mostly soil). They excrete oxygen (their gaseous waste product) into the atmosphere and store sugars and proteins in their tissues.

Herbivorous animals (humans included) take advantage of this bonanza of resources, inhaling oxygen and eating plants to fulfill nutritional requirements. Animals then routinely excrete solid and liquid wastes into the environment (future plant nutrients) and exhale carbon dioxide (our gaseous waste product) into the atmosphere, providing photosynthetic plants the opportunity to continue the cycle of life. When plants and animals die, as they all must, communities of soil-based microbes known as detritivores return the elements contained in their carcasses to the earth by the process of decay, providing a kind of natural fertilizer for the next generation of plants; it's a natural "ashes to ashes" strategy for nutrient recycling. It has existed this way for some 400 million years and will undoubtedly go on for some time to come, with or without us. The fact that it has survived for so long in the face of extraordinary environmental changes suggests strongly that it is an incredibly resilient and highly redundant system, one that is almost impossible to destroy. This augurs well for the ability of fragmented ecosystems to repair themselves if we simply learn how to keep our hands off and mind our own business.

STRENGTH IN NUMBERS

When a mixed group of plant species, all with similar tolerances for temperature and humidity, grow in a given geographic region, their very presence attracts animals of different species to coinhabit that region. The result is the eventual establishment of mutually dependent relationships, in which all the life forms in that zone, including the microbes, join to share in the flow of energy provided by the sun. This is the bare-bones definition of a functional ecosystem. Ecosystems are also known as biomes. Mostly, ecosystems refer to terrestrial situations, and for our purposes, I will stick to this definition. The one characteristic they all share is that primary productivity (the total mass of plants produced over a year in a given geographically defined region) is limited by the total amount of energy received and processed. In fact, the amount of available energy actually determines the very nature of each ecosystem. For example, rain forests have an abundance of sunlight and a year-round growing season, allowing all of the inhabitants that live there to prosper. In contrast, alpine forests are limited by a short growing season and lack of warmth. No ecosystem can exceed the limits of biomass production, which is strictly regulated by the total amount of incoming energy, period. In years of high productivity, energy is used to its maximum efficiency, and in lean years, largely regulated by fluctuations in weather patterns, the result is lower bioproductivity. Nature adjusts to a varying supply of calories. Cities do not follow this simple rule of nature, and therein lies the problem.

VIVE LA DIFFÉRENCE

Ecosystems vary from place to place, from the kinds of plants and animals found in each to the physical makeup of the landscapes. The most important features of an ecosystem are the annual temperature regimes and precipitation profiles, which vary greatly with latitude and altitude. Hence, there is a plethora of varied, vibrant, robust assemblages of life that have flourished for hundreds of thousands of years. Only recently in geological time have we been able to make any impact on their functionality. In just the last ten thousand years we have spread ourselves over the entire planet, encroaching into all terrestrial ecosystems and fragmenting most of them with our farms, grazing lands, and human settlements. We invented agriculture at least six different times across the entire globe. Food production freed us from wandering and allowed for the rise of what we have come to refer to as civilization. Unfortunately, along the way we forgot to pay attention to the processes that encouraged our own evolution—processes that are still at work today. Many ecologists, myself included, hold that unless we make peace with the natural world, we will surely lose our place in it.

THE ENEMY WITHIN

To frame the problem in an ecological perspective, in stark contrast to the natural world around us, urban centers (the "technosphere" described by William McDonough and Michael Braungart in *Cradle to Cradle*) have no apparent cutoffs regarding constraints of growth. This is especially true in the poorest countries. It's a rare situation that results in uncontrolled growth due to extreme wealth, but it happens, as well. Abu Dhabi, Dubai, and the United States routinely exceed their quotas for almost every resource, including food, water, and energy. The result of such excessive behavior has led to the problems facing us today. By defining the problem in ecological terms, we may be able to pave the way for a complete overhaul of the way we carry out our daily lives. Today, nearly 50 percent of us choose to live in cities and surrounding suburbs. These crowded urban centers rely heavily on importing food, ores, and other essential resources. If we continue to rely on harvesting resources from an environment we have created, whose production is solely dependent on using more and more fertilizers, herbicides, and pesticides, those forced ecological situations will soon fail and we will be left stranded. In fact, many agricultural regions are already failing, and others are soon to follow.

So, the real question is, can a city bio-mimic an intact ecosystem with respect to the allocation and use of essential resources and, at the same time, provide a healthy, nurturing, sustainable environment for its inhabitants? As the reader will see in what follows, I think the answer is an emphatic yes. In fact, we have no choice if we want not just to survive but to thrive. We have all the tools to do so. All we have to do is apply them creatively to address this single question. Built into this ecological survival strategy is the eventual repair of much of what we have damaged along the way to becoming seven billion strong.

HAVING IT BOTH WAYS

Repairing the environment and still having enough to eat may seem like mutually exclusive goals. If the world's population continues to increase and we need to place more and more land into agriculture, and if in doing so we are forced to cut down more forest, how can we expect the environment to heal itself? In theory, the solution is straightforward: Grow most of our food crops within specially constructed buildings located inside the city limits using methods that do not require soil. This would allow for the conversion of an equivalent amount of farmland back into whatever ecosystem was there originally, usually hardwood forest. The regrowth of the forests would eventually sequester significant amounts of carbon from the atmosphere and begin the healing process. Biodiversity would be increased, and ecosystem services, such as flood control and cleaning of the air, would be strengthened. The more urban farms there are, the larger the amount of carbon that would be converted to cellulose in the form of trees. It is that simple.

TECHNOLOGY RULES

To most who hear about this scheme for the first time, it all sounds too simplistic to actually have any chance of working. It sounds downright naive and impractical. Yet, over the last ten years, the more I and my 106 bright and enthusiastic graduate students thought it through, the more reasonable the idea became. We called it "vertical farming." It is a concept whose premise is easy to envision: Stack up "high-tech" greenhouses on top of each other and locate these "super" indoor farms inside the urban landscape, close to where most of us have chosen to live. However, I came to realize early on that making it happen will not be an easily attainable goal, and certainly not simple from an engineering and design perspective.

Although there are at present no examples of vertical farms, we know how to proceed—we can apply hydroponic and aeroponic farming methodologies in a multistory building and create the world's first vertical farms. Some parts of the world are rapidly moving toward such a scheme already, especially those countries—the Netherlands, Belgium, Germany, Iceland, New Zealand, Australia, China, Dubai, Abu Dhabi, and Japan, to name but a few—that are running short of arable farmland and have the resources to contemplate replacing the accepted traditional agricultural paradigm with something new and more efficient. Other, less affluent countries, such as Niger, Chad, Mali, Ethiopia, Darfur, and North Korea, desperately need vertical farms to rescue enormous populations from extreme hunger.

Vertical farming practiced on a large scale in urban centers holds the promise that sustainable urban life is not only possible but highly desirable and technologically achievable. With all the advances made over the last ten years in the sustainable use of resources, a city can now choose to become a functional urban equivalent to a natural ecosystem by employing high-tech versions of waste-to-energy strategies, food production, and water-recovery systems. In that way, it can process all resources that generate waste back into usable resources without further damaging the environment.

DESIRABLE ATTRIBUTES

Ideally, vertical farms should be cheap to build, modular, durable, easily maintained, and safe to operate. They should also be independent of economic subsidies and outside support once they are up and running, which means they should also generate income for the owners. If these conditions are realized through an ongoing, comprehensive research program that leads to construction of efficient, productive vertical farms, urban agriculture could provide a continuous, abundant, and varied food supply for the 60 percent of the population that will live in cities twenty years from now. Ironically, the migration to cities is being driven by the "plight" of the farmer. People move to cities for economic reasons—when a city's economy is good it pulls people to it. Droughts and floods that affect huge areas of agricultural land result in mass migration of farmers to cities in bad times. Urban farming opportunities that arise directly from the creation of vertical farms could provide jobs for these people. What could be a better outcome for displaced agricultural personnel than for them to discover that they can still plant and harvest, only now in a controlled environment? No more praying for rain or sunshine or moderate temperatures; they could save their prayers for things like winning the lottery.

CONTROL FREAK

Farming indoors is not a new concept; greenhouse-based hydroponic agriculture has been in existence since the 1930s. Numerous commercially viable crops such as strawberries, tomatoes, peppers, cucumbers, herbs, and a wide variety of spices have seen their way from commercial greenhouses to the world's supermarkets in ever-increasing amounts over the last fifteen years. Most of these operations are small by comparison to traditional soil-based farms, but unlike their outdoor counterparts, these facilities can produce crops year-round. Sweden, Norway, the Netherlands, Denmark, England, Germany, New Zealand, the United States, Canada, Japan, South Korea, Australia, Mexico, Spain, and China all have thriving greenhouse industries. In addition to plants, some animal species have been commercialized by indoor farming, including freshwater fish (tilapia, trout, striped bass, catfish, carp) and a wide variety of crustaceans and mollusks (shrimp, crayfish, mussels). Cattle, horses, sheep, goats, and other large farm animals seem to fall well outside the paradigm of urban farming. However, raising fowl (chickens, ducks, geese) and even pigs is well within the capabilities of indoor farming.

Vertical farming promises to eliminate external natural forces as confounding elements in the production of food. Much of what we plant never gets a chance to grow to maturity due to adverse weather events driven by rapid climate changes that are, in turn, linked to an ever-increasing rate of CO_2 emissions. Today, the United States Department of Agriculture (USDA) estimates that over 50 percent of all crops planted in the United States never reach the plate of the consumer. Droughts, floods, spoilage, and plant diseases account for most of the losses. On a worldwide basis, the situation is even worse, with nearly 70 percent of planted crops never reaching the harvest stage, succumbing—in addition to the things listed above—to attack from insect pests such as locusts and a wide variety of endemic microbial pathogens. These losses are totally avoidable, since we can now grow most of what we need to eat inside under carefully selected and well-monitored conditions that ensure an optimal yield for each species of plant and animal year-round. The choice is simple: Control everything (indoor farming) or control nothing (outdoor farming).

Today, we stand at an interesting, albeit daunting, crossroad. We continue to urbanize without incorporating the necessary skills to live sustainably, and struggle to understand enough about the damaging effects our penchant for consuming everything in sight is having on ecological processes. In this regard, science has led the way, with satellites reporting on many of the factors that contribute to our present dilemma of rapid climate change.

DO NO HARM

On a global scale, we need to emulate the physician's credo: "Do no harm." In this case, "do no harm" means helping the rest of life on earth to survive. In doing so, we help ourselves, as well. On the other hand, some tend to ignore the long-term consequences of their actions and opt for an immediate return on investment. In many cases, this results in different forms of encroachment into natural systems, disrupting ecosystem functions and services and eliciting a host of health problems that were clearly avoidable.

WE ALL LIVE DOWNSTREAM

One of the most pressing reasons to consider converting to urban agriculture relates to how we currently view and handle agricultural waste. In fact, we don't handle it at all. Agricultural runoff is responsible for more ecosystem disruption than any other single kind of pollution. Most of the world's estuaries have been so adversely affected by runoff that they no longer function as nurseries for the ocean's marine fish, crustacea, and mollusks. That is why the United States must import more than 80 percent of its seafood from abroad. What's more, we can do nothing about it in most instances, since floods dictate the timing and extent of the runoff events. Vertical farms would recycle their own water, thereby eliminating agricultural runoff once and for all.

WATER, WATER, EVERYWHERE

Liquid municipal waste (black water) is handled differently from solid municipal wastes, such as cardboard. Most often, in less developed countries, grey water and even black water are flushed without treatment. This greatly increases the risk of people contracting salmonella, cholera, amoebic dysentery, and other infectious diseases transmitted by fecal contamination. Instead of getting rid of the waste altogether, ideally, one would want to capture the energy in human fecal solids. A gram of feces, when incinerated, yields some 1.5 kilocalories of energy. If New York's 8 million citizens decided to pool their fecal resource and generate electricity by incinerating it, they could realize an astounding 900 million kilowatts of electricity per year. That's enough energy to provide electricity for many large versions of a vertical farm without tapping into the municipal grid.

Some vertical farms will be engineered as stand-alone water-regenerating facilities. They will take in safe-to-use grey water and restore it to drinking-water quality by collecting the water of transpiration using advanced dehumidifier systems. Harvesting water generated from transpiration will be possible because the entire farm will be enclosed. The resulting purified water will then be used in other vertical farms to grow commercial crops and for aquaculture.

Ultimately, any water source that emerges from the vertical farm should be drinkable, thus completely recycling it back into the community that brought it to the farm to begin with. Again, using New York City as the example, the "Big Apple" discards some 1 billion gallons of treated grey water every day into the Hudson River estuary. At a conservative five cents a gallon for industrial-quality water, it appears to be well worth the effort to recover it.

DUST TO DUST

Another major source of organic waste comes from the restaurant industry. In New York City there are more than twenty-eight thousand food service establishments, all of which produce significant quantities of "leftovers," and the restaurateurs pay a hefty price to have it carted off. Stacks of extra-heavy-duty garbage bags bursting at the seams with the stuff sit out on the curb, sometimes for hours to days prior to collection. This allows time for cockroaches, rats, mice, and other vermin to dine al fresco at some of the finest restaurants in the Western Hemisphere. Vertical farming may allow restaurants to be paid (perhaps according to the caloric content) for this valuable commodity. Not only would an industry with a notoriously small (2–5 percent) profit margin earn additional income, it would provide raw material to be recycled through waste-to-energy schemes. Oh, and one more thing: Good-bye, vermin. In New York City, there are eighty to ninety restaurant closings each year, the vast majority of them precipitated by inspections conducted by the Department of Health. A common finding by inspectors is vermin (mouse and rat droppings, cockroaches) and generally unsanitary conditions that encourage the persistence of these unwanted diners. Eliminating significant populations of vermin by controlling the amount of restaurant waste left out on curbs and inside kitchens could encourage the development of abandoned inner-city properties for middle- and low-income housing, and without the health hazards associated with having to share space with the four- and six-legged variety of tenants.

EFFLUENT SOCIETY

However great the contribution of urban waste is to the destruction of terrestrial and aquatic ecosystems, it is agricultural runoff that wins the gold medal for pollution worldwide. As already alluded to, farm runoff despoils vast amounts of surface water and groundwater. Some 70 percent of all the available freshwater on earth is used for irrigation, and the resulting runoff, typically laden with leftover salts, herbicides, fungicides, pesticides, and fertilizers leached from the nutrient-depleted farmed soil, is returned to countless rivers and streams. Runoff that reaches the oceans untreated has the potential to disconnect other ecological systems through its nutrient-loading and oxygen-scavenging agrochemicals, particularly nitrates and nitrites. Estuaries and coral reefs have been the hardest hit. For example, agricultural runoff from farms in Jamaica has reduced the coral reefs in the surrounding ocean to nearly barren remnants of once abundant undersea life. This, in turn, has shut down a fishing industry that was wholly dependent on the intact coral reef for its existence. Similar results from deforestation for purposes of creating farmland have forever altered the reefs surrounding Madagascar. And a major flood in 1993 along the middle reaches of the Mississippi River left the ocean life of the Gulf of Mexico reeling for years afterward. A dead zone caused by the mobilization of nitrates left in the soil from years of agriculture along the fertile bottomland of that river system killed off an entire fishery (oysters, shrimp, fish) from Port Arthur, Louisiana, to Brownsville, Texas. Hurricane Katrina delivered the latest blow that will most likely ensure that this once productive coastal fishery remains in the dead zone for decades to come.

SPACED OUT

Vertical farming offers the possibility of greatly reducing the quantity of this nonpoint source of water pollution. The concept of sustainability will be realized through the valuing of waste as a commodity. We are now able to live for long periods of time in closed systems (e.g., the International Space Station) off the surface of the Earth, and in that instance, the concept of waste is already an outdated paradigm. Unfortunately, this goal has yet to be fully realized, even by NASA. So if we are to live continuously on the moon or Mars, then we had better learn how to do it here first. I will offer the reader my views on how we might proceed to the first vertical farm, but I have no doubt that others are working hard on the creation of a practical version of one, as well. May we benefit from everyone's efforts and enter into the next phase of our evolution with a greater sense of security about the essentials of life itself—a safe and constant source of food and water.

CIRCULAR REASONING

To emulate the behavior of an ecosystem means to live within our means with regard to recycling energy, water, and food, and in dealing in a realistic and responsible fashion with populations. Most important, we must learn to handle the problem of waste management ecologically. In nature, there is no waste. When seen through the eyes of the ecologist, the city fails to meet even the minimum standards of the simplest of ecosystems. Everything that the city consumes comes from outside its limits: energy, water, food, dry goods. Add to that the millions to billions of dollars that are spent annually trying to get rid of waste, and you end up with a crazy-quilt system that works exactly opposite to the way we would have designed one a hundred years ago if we knew what pitfalls lay ahead.

The main premise of this book is to focus on food grown inside tall buildings within the cityscape, but if we can learn to do that, then we can also figure out what to do with the waste generated by vertical farming. Solving that problem (which would require no new technologies) would solve all the other waste-management problems, too. The bio-mimic principle has grown recently and is now the mantra for Silicon Valley and other regions of the techno-sphere. The logic system (i.e., copying what nature does best) that spawned the nanotechnology industry has led virally to a host of related new companies, and will continue to grow as we learn more about how nature has solved its problems of coping in an ever-changing environment. Howard Odum, the noted ecologist, once remarked: "Nature has all the answers. What is your question?" Mine is, how can a city bio-mimic a functional ecosystem?

CHAPTER 2

YESTERDAY'S AGRICULTURE

If we do not change our direction,

we are likely to end up where we are headed.

—CHINESE PROVERB

A MOVABLE FEAST

Sometime in the year 2010, the human population exceeded seven billion. This is more than a little disconcerting. The World Health Organization and the Population Council estimate that, given the current rate, by the year 2050, we will top out at around 8.6 billion. When I was born, there were only 2.6 billion of us. As if that were not enough to be concerned about, the human population is not the only thing on the rise. Our planet is developing a "fever," an obvious indication that something is wrong with the entire system. Global warming, also referred to as climate change, is an unanticipated consequence of the unprecedented growth in our population. The ice is melting all over the globe: Earth is suffering from a colossal case of a "bipolar" disorder. It's directly linked to our penchant for using more and more fossil fuels to accommodate our increasing demand for food and manufactured goods. If we continue with our current food-producing strategies, getting enough high-quality, safe produce to 8.5 billion people will define the next crisis we must address and remedy if the human species is to survive. How did things get so out of control? To answer this question, we have to understand how agriculture arose to begin with. Along with that invention, it is necessary to document how we managed to escape an early extinction and emerge into the light of the modern era as the "masters" of our small part of the universe.

Looking back to our origins on the verdant, fertile plains of East Africa, and our rise to prominence as a dominant mammalian species, it all seems quite improbable. Three million years ago, all of the hominid species (just how many we still are not certain) were slow moving compared to even the laziest large cats, of which there were many kinds. Fight or flight was not an option. Avoiding those predators was key. To do so, our forerunners had to rely largely on cunning and guile rather than on physical prowess, or else be eaten. A handful of plausible theories may explain how our ancestors managed to avoid elimination, all supported by careful examination of the fossil record. One group of physical anthropologists contends that human evolution was favored over the strictly herbivorous hominids due to our ancestors' high-protein diet that included meat, enabling them to rapidly develop a larger brain. That, coupled with the development of opposable thumbs, allowed humans to acquire a remarkable level of dexterity that, in turn, enabled them to invent and manufacture tools, especially weapons. As an aside, it is ironic that once humans were the only hominid species left on earth, much of the advancement of the species, right up to the present, was dependent upon the development of superior weaponry. Advanced techniques for hunting, as well as for defense against predators and other competing hominid species (such as those portrayed by Stanley Kubrick in the opening scene of the film version of Arthur C. Clarke's riveting novel *2001: A Space Odyssey*) were adopted over less efficient gathering and scavenging strategies. This supposedly allowed humans the freedom to track and hunt the migrating herds of gnus, zebras, and other large grazing herbivores.

A Nonmovable Feast

A far less popular but equally plausible theory holds that, instead of making tools for the purposes of hunting and expressing our technological superiority, our ancestors functioned in those savannas as opportunistic omnivores whose main activity was scavenging. Humans used hand axes and other simple tools to break open the long bones of abandoned carcasses of animals that had been recently killed and completely stripped bare of flesh by lions, tigers, leopards, vultures, and the like. The nutrient-rich marrow might have been all that was left for them to consume, but it would have been enough of an energy and high-protein hit, supplemented with the local edible fruits, nuts, and grains, to allow them to flourish in an otherwise hostile environment. The notable absence of hyenalike ancestors in Africa, but not in western Europe, during early human development further strengthens this hypothesis, since the present-day hyena, a notably aggressive species, is the only East African predator/scavenger capable of cracking open bovine long bones with its incredibly strong jaws and durable teeth. In fact, they can make fast work of elephant and hippo bones, too. Their presence in humanity's original habitat would undoubtedly have negated any opportunity afforded us to harvest marrow, especially in the absence of any kind of effective weapons such as spears. Perhaps it is best to view the level of success our ancestors enjoyed during their early development as the deployment of a combination of survival strategies, a sort of "whatever works best" approach, applied on the spot at the time of need. Though most anthropologists rarely agree on anything related to human evolution, they all seem to believe that, if nothing else, early hominids were resourceful, thoughtful mammals capable of a wide range of behaviors, grounded in a strong, instinctual desire to survive long enough to reproduce. Early populations of modern humans (who lived about two hundred thousand years ago) can be easily identified as omnivorous hunter-gatherers by examining the grinding patterns on their fossilized teeth—clear evidence that they ate whatever was most available and safe to harvest. The marks left on many broken animal bones found at gathering sites give indisputable evidence that humans cracked open those bones with hand axes. For many thousands of years, they managed to eke out a living this way without a thought to growing edible plants or settling down and urbanizing.

And Then There Was One

Neanderthals were one of the last remaining hominid species still around after humans evolved into their final genetic signature of *Homo sapiens sapiens* (*Homo floresiensis* aside). While *Homo sapiens sapiens* was trying to figure out how to leave South Africa, these close relatives had escaped earlier, around 250,000 years beforehand, as *Homo helmei,* and migrated northward into Europe. There they evolved into *Homo neanderthalensis* and spread out, occupying most of Eurasia beginning around 130,000 years ago. They crafted a wonderful set of tools, mostly heavy spears for jabbing and thrusting but probably not for throwing long distances, with which they hunted, skinned, and carved up their game animals.

Their main quarry were the mammals of the Pleistocene megafauna, including wooly mammoths and cave bears. They remained attached to a hunting-and-gathering lifestyle that died out almost the moment *Homo sapiens sapiens* arrived on the European scene some forty-five thousand years ago. Apparently a few humans were attracted to Neanderthals, since our genome has recently been shown to contain about 2 to 4 percent DNA sequences from the Neanderthal genome (which has recently been sequenced), proving that the two species produced hybrids.

By twenty-eight thousand years ago, mysteriously, there was not a single Neanderthal to be found anywhere. How and why they disappeared so fast is the subject of much speculation, including the possibility that humans merely applied their superior weaponry to the situation and eliminated the competition. While there is scant but convincing evidence that the two species of hominids did have the occasional armed conflict, I personally don't believe that humans "whipped their butts." Neanderthals were excellent hunters and banded together. Defeating such a strong enemy time and again would have challenged even the most clever cohort of humans. It is much more likely that diseases did them in; humans', to be precise. Mitochondrial DNA extracted from Neanderthal bone samples of a total of six individuals gathered from sites in Spain, Germany, Croatia, and Russia show that their own genome was remarkably homogeneous, only one-third as diverse as modern humans. This suggests that the Neanderthal's immune system was also highly restricted. They would have been well suited to fight off microbial infections that they alone had encountered and evolved with, but the introduction of new microbial pathogens with different antigenic signatures and more powerful virulence factors from encroaching human populations out of Africa would have taken the repertoire of Neanderthal T and B cells by complete surprise. The result might well have led to the extinction of this last remnant hominid species, clearing the way for humans to repopulate that region, which they did.

STRANGERS IN A STRANGE LAND

It would not be the first time newly introduced infectious agents caused the extinction of a species. The natural history of the Hawaiian Islands is rich with examples of nonindigenous infectious agents that wiped out at least five genera of tropical birds through the introduction of bird malaria. The same kind of thing almost happened when the Spanish invaded South and Central America and unintentionally killed off nearly 90 percent of the 50 million native people there. The conquistadores exposed these unfortunate innocents to unfamiliar microbes, including influenza and the common cold, they brought with them to the New World from Europe. The result was near mass extinction. Before the Spanish gave up on their dream to establish new colonies in that part of the world, nearly 45 million people had perished. In exchange, the Spanish troops received the lasting "gift" of syphilis from indigenous populations, undoubtedly acquired through raping and pillaging sorties, which they then introduced into Europe. Syphilis is not as fulminating a disease as influenza, and therefore, Europeans died more slowly than the natives had, mainly from the neurological form of the infection. Since it is a sexually transmitted bacterial infection, many Europeans simply avoided syphilis by remaining faithful to their family unit.

HUNT, GATHER, SLEEP. HUNT, GATHER, . . .

One thing is quite certain: Neanderthals were never farmers. The climate in most of northern and central Europe and eastern Asia did not favor an agrarian way of life. For one, the growing seasons were short and the availability of arable land was scarce. Second, there was no environmental need to invent farming, since they were cave-dwelling hominids who followed game animals during their migrations and lived off the meat well into the winter months by cleverly taking advantage of the freezing conditions outside to cache and preserve their food. Neanderthals did, however, collect wild grains and other edible plants from their immediate environment to tide them over through the lean times when game was in short supply. A portion of this natural harvest was bound to grow in greater abundance near where they lived, as some seeds—dropped accidentally in transit from collection site to cave or blown from stockpiles into the adjacent landscape—would have taken root in nearby soil. Nonetheless, despite this windfall profit of benign carelessness, Neanderthals did not cotton to the idea that growing plants for food was a good idea. When humans began to live in similar ways later on, it eventually "clicked" that by purposely or even accidentally dispersing some of the collected seeds into nearby fields next to water sources (i.e., environments in which these valued plants already grew), they could create a more reliable source of food.

IN THE BEGINNING

A recent archeological excavation led by Ian Kuijt and Bill Finlayson in Dhra, Jordan, near the Dead Sea, has shed new light on the origins of agriculture in that region. They uncovered a storage bin-like structure with the remnants of barley seeds and grindstones dating back to around 11,300–11,175

BP. These early farmers apparently collected wild seeds, grew them locally, and then used specifically designed tools for processing them into flour. The fact that they needed storage bins strongly indicates that they were highly successful agriculturists. This site predates by at least one thousand years the first-known site for farming domesticated grains.

Another discovery in that same region dates the domestication of barley to around nine thousand years ago. Farming, especially then, was hit or miss. When to plant and harvest had to be worked out for the very first time, and without the aid of an extension agent from the USDA or the Internet. Timing is everything. Enter the invention of the calendar, astronomy, mathematics, written language, and, last but certainly not least, religion. But this is getting a little ahead of the story.

ON THE ROAD

Life was looking pretty good up until about one hundred thousand years ago, when our ancestors were apparently forced out of East Africa by changes in climate that dried out the forests, turning them into semiarid grasslands and savannas. Perhaps this is also when hyenas entered the scene from an ice-bound Europe. Whatever the reasons, the results were clear enough. Humans first wended their way southward into what is now the Cape Town region of South Africa and hunkered down in caves up along the west coast, somewhere between what is now Namibia and Angola. The hyenas stayed the course in the Serengeti and settled into their present lifestyle, since their main source of food, the game of Africa, remained confined to that region of East Africa. Then, some twenty thousand years later, as ocean levels began to drop even more (around 400 feet) due to the onset of yet another ice age, humans somehow managed to escape out of Africa, perhaps by walking back up along the east coast to the Middle East, then to South Asia, into Asia, and eventually across the Bering Strait land bridge to the North American continent. In many places, they stayed and created permanent settlements, as harvesting local plants and hunting the readily available local game animals gave them a lifestyle they could settle down with and enjoy. In most centers where the invention of farming occurred, settlement of the region happened first, then farming arose, not the other way around, as logic might dictate. In other words, when humans finally established large gatherings of extended families, they apparently felt the strong pull of settlement life, with its many social advantages, and created a way to stay put and at the same time to accommodate a growing population that, without crops, might have exceeded the limits of the environment to feed. In fact, many early settlements did fail due to the lack of a reliable food supply, such as grains, that could be stored and used at a later time. Farming seemed like a natural outgrowth of that desire to remain connected. Humans are by nature gregarious. No wonder, then, that farming, once it developed, took hold in so many geographic locations. This all came about around ten thousand to twelve thousand years ago.

GREAT MINDS THINK ALIKE

There were six major regions of the world in which agriculture arose and then spread to adjacent areas over several thousand years: the Near Eastern Center, the Central American Center, the Chinese Center, the New Guinea Center, the South American Center, and the North American Center. Semidesert regions were favored sites for the invention of farming. Again, this flies in the face of logic. The most plausible explanation is that agriculture could be more successful in the absence of competitor bands of humans and large four-legged predators. The Anasazi in the Southwestern region of North America; the Nazca, Wari, and Incan peoples in South America; and of course the Hittites, Babylonians, Sumerians, and Egyptians all produced stable settlements, and many of them flourished for hundreds to thousands of years with farming as the basis for sustaining them.

GOOD-BYE BIOSPHERE, HELLO TECHNOSPHERE

Once farming became routine and reasonably predictable, we proceeded to convert much of the earth's natural landscape into food production. History has recorded in a wealth of cultural expressions the progression of events regarding the evolution of settlements and cities; the emergence, flourishing, and eventual collapse of entire civilizations; and especially the relentless, irresistible growth of the human population. In the process, we systematically fragmented most of the world's terrestrial biomes, rearranging the lives of countless assemblages of plants and animals and causing the extinction of many others. Ultimately, because all life on earth is connected in some way or another, even we became victims of our own penchant for severely altering natural systems. The loss of ecosystem services was the result. We found out the hard way just what nature provides for us, and free of charge: flood control, purification of the air, and regeneration of freshwater, not to mention the regulation of the earth's temperature. In fact, the earth has been changed so much by our drive to commit more land to food production that, no matter where we look, there is abundant evidence of extensive ecological damage.

Technologies for sustaining food production from one

year to the next—designing of irrigation schemes, food processing and storage systems, complex cuisines, development of astronomy and calendars to be able to predict the seasons, written languages, and organized religions—were in their earliest phases of development in many places around the world ten thousand years ago. All of these embryonic human urban centers were pulled together helter-skelter, ushering in the first of two major agricultural revolutions. However, farming arose somewhat more gradually than is commonly depicted in the history books.

As alluded to earlier, Neolithic communities at least two hundred thousand years ago were routinely harvesting and processing a wide variety of wild and abundant grains, including founder species for all major crops such as wheat, corn, millet, barley, and rice. There is ample evidence for this in numerous cave sites through Europe, Asia, and even South America. Most anthropologists believe that seasonal shortages of wild edible plants eventually led to the systematic management of these forerunner crops as a strategy to avoid overharvesting. Four examples will illustrate how pervasive and remarkably similar the development and practice of farming was, regardless of location or crops selected. It is almost as if someone had simultaneously turned on a switch in the "I want to be a farmer" part of our genome.

THE MIDDLE EAST EXPERIENCE

Around seven thousand to eight thousand years ago, in the region dubbed the Fertile Crescent in modern Iraq, early attempts at growing food in large amounts in a sustainable fashion met with limited success, at best. That is because arable land was as scarce then as it is today, only occurring on the floodplains next to rivers such as the Tigris and Euphrates. Settlements throughout the region had access to wild grains and cereals such as barley, at least three kinds of wheat, and legumes such as chickpeas, beans, and lentils. As farmers gathered up these grains and seeds, then cultivated them under irrigated conditions, these founder crops grew weaker with respect to their ability to survive without our help, and at the same time became more nutritious, since the largest sizes of grains and seeds were selected for the next year's planting. This is the process of artificial selection in its earliest practice. A few wry observers of the human condition have said that what actually happened was that wild plants in fact cultivated *us* by trapping humans into becoming totally dependent upon them for their very survival, thus ensuring the plants' own survival in the process! This sort of a "you feed me and I'll feed you" hypothesis of symbiotic relationships is an interesting twist on the practice of farming, to say the least.

WASTREL SOCIETY

But without any awareness of the need to practice methods that conserved the soil from year to year, the earth's first farmers unwittingly allowed their crops to consume valuable nutrients without a thought to replacing them, and in doing so damaged the land beyond repair. Of course, these were the earliest of days. How could anyone possibly begin to understand the physiological needs of plants? Most believed that the gods—Tammuz and Nissaba (Babylonian), Osiris (Egyptian), Demeter (Greek), Saturn (Roman), Mama Allpa (Incan), Kukulcan (Mayan), Kokopelli (Anasazi), Hou Chi (Chinese)—were in charge of whether or not the crop would be good. Agricultural failures gave rise to whole religions; at the center of many was the belief that the community or some subset of individuals in that community had done something to offend those deities. To "correct" the problem and guarantee the next year's harvest, sacrifices were often made, some of them at the cost of human life. Today, most farmers still pray for the essentials of what goes into reaping a bumper crop—just the right amount of rain, moderate temperatures, and lots of sunshine. In ancient times, as crops eventually failed due to lack of knowledge regarding anything to do with sustainable land use practices, farmers of the Fertile Crescent soon transformed that area into a desert by continuing to move their operations northward until they eventually ran out of floodplain altogether. That particular region of the Middle East still suffers from prolonged drought and the lack of nutrient-laden floods, thereby severely limiting any form of agriculture along those two famous river systems.

EGYPT

In contrast, land along the Nile sustained farming throughout the millennia due to climatic conditions that ensured a more stable hydrological cycle associated with periodic flooding events. At various times throughout Egyptian history, farmers cultivated an incredibly varied array of vegetables and fruits, including garlic, leeks, onions, cabbage, lettuce, cucumbers, radishes, asparagus, legumes (peas, lentils, beans), olives, dates, and many herbs and spices. Many of the latter were used for medicinal purposes, as well. Nonetheless, even there devastating droughts produced crop failures, many of which are accurately recorded in ancient papyri and hieroglyphs. Egyptians even had a god assigned to watch over the growth of the vegetation (crops) and the weather, Osiris. They were superb observers of the natural world around them and drew heavily upon it for their belief systems. A good example is the lowly dung beetle. Also known as scarab beetles, these abundant insects were worshiped as the very givers of life itself. The beetles occupied a peri-domestic niche, dependent upon domesticated animals for a life-giving substance—dung. The life of this small insect was an open book, save for what happened underground. Many an Egyptian citizen saw them taking portions of common animal droppings, rolling them into a perfect sphere, and then, with their back legs, shoving the sphere down a hole they had dug. The next spring, out of the hole "miraculously" emerged a sprouting seedling, usually some kind of grass, and another beetle. How remarkable that life itself should arise from the very stuff that we discard after consuming life. To the Egyptian scholar, this sequence immediately suggested a mutually dependent cycle of sorts. Of course, the sun (Ra) was also deified, so the Egyptians may well have been the first ecologically savvy civilization, recognizing all the important connections needed to sustain life. They also invented a remarkable system of irrigation schemes to bring water from the Nile inland many miles, thereby extending their farming operations. Millions were fed from the grains and other produce carefully tended by a dedicated cadre of professional farmers.

Today, the flow of the Nile is severely impeded by the high dam at Aswan, which prevents the entire river basin from flooding and replenishing its banks with nutrient-rich silt once culled from the steppes of Sudan and Ethiopia during the rainy season up country. Modern irrigation projects and the extensive use of fertilizer have compensated for the loss and allowed farming to continue unabated in that desert region.

One other negative unintended consequence of the high dam has been the increase in the geographic range of schistosomes—waterborne parasites associated with agricultural practices in which human feces and urine are used as fertilizers. Once restricted to the lower Nile, the parasite's intermediate host, a snail, has been able to extend its range to the foot of the dam itself, thanks to the greatly reduced flow of the river. This parasite group causes life-threatening diseases that still persist in that region today, despite Egypt's institution of sound public health practices aimed at its eradication.

The ecological damage incurred by farming in both the Middle East and Egypt has been extensive, and in the case of the Nile, preventing flooding has adversely affected even the sea life of the entire Mediterranean basin. Other regions have suffered similar fates for the same reason. In many cases, the result of ecological collapse due to poor farming methods has led to the fragmentation of cultures, or their extinction.

MEXICO

Mexico is a major hot spot for the birth of agriculture. The origins of modern corn can be traced back to Balsas Valley in the south-central portion of that country. In 2009 agroarcheologists found strong evidence in the form of microscopic characteristically shaped starch granules at various sites in that region that pinpoint the origin some 8,700 years ago of the cultivation of maize, the parent precursor of all modern varieties of corn. Maize arose as the result of several related grasses hybridizing in nature to produce a plant that allowed for its cultivation in semiarid conditions.

Today's corn is a keystone crop for many countries, including the United States. In 2007 the United States exported 63 million metric tons of corn, representing some 64 percent of the world market. That's amazing when we consider that a mere eight thousand years ago we were just learning how to grow the stuff. Soon after its cultivation, maize became a staple food crop in many places throughout the New World, spreading northward to the Anasazi in the American Southwest and southward into South America to the Nazca, Wari, and later the Incas. More than 50 percent of the diet in most of these New World cultures consisted of maize and products derived from that crop, such as beer.

PERU

Two major cultures arose in Peru: the Nazca (circa 100–300 BP) and the Inca (1200–1500 BP). Both lived in regions of the world that were, to say the least, parched environments: In some parts of the Peruvian Alticama, it hasn't rained even once in over two thousand years. Both responded by creating innovative irrigation systems, some of which are still in use today. Agriculture was more than a challenge, since the soils were poor due to the lack of regular precipitation events. Nonetheless, both areas had robust cultures and complex cuisines. The Nazca were famous for their incredibly large, accurate, naturalistic, and abstract figures they "drew" in the desert by arranging stones to depict the images. Discovered in the 1920s, for many years the purpose(s) of the Nazca figures remained a total mystery to even the most persevering teams of archeologists. That is, until recently, when it became apparent after extensive computer simulations that they were not used as astronomical aids, as originally hypothesized. Instead, the most logical explanation seems to fit their geographic orientation: Most of the figures have at least one feature that points to a source of underground water. The Nazca were so dependent upon knowing where to get their water for irrigation that they conceived, then executed, these remarkable figures. Having a reliable source of food forced early humans to use every fiber of their thinking process, or else they perished. It is that simple.

The Inca evolved an equally complex society to that of the Nazca, and, prior to the arrival of the Spanish, lived in splendid semi-isolation in the mountainous regions to the east of the center for the Nazca civilization. There they cultivated a wide variety of edible plants, including tubers such as potatoes, which they could store by first drying them and then grinding them into a powder. They used the potato powder to make breadlike food items. Of course, they also grew maize, which they acquired by trading with cultures from the north. In addition, several kinds of tomatoes and peppers, avocados, fruit (strawberries and pineapple), nuts, and even chocolate were staples of the Incan diet. To grow them, they constructed elaborate high-altitude irrigation canals that brought water from great distances to places as remote as Machu Picchu. Many of these early irrigation projects were so well constructed, they remain in use today.

INTENDED CONSEQUENCES

Coincident with the advent of agriculture was the development of texturally rich written languages. The origins of spoken language arose early in our history, perhaps two hundred thousand years earlier than writing. It was writing, however, that enabled us to leave a record of what to do and what to avoid doing. Planting schedules, what to plant, how to plant, where to plant, when to harvest, how to store grain, and so forth all had to be written down. Equally important was knowing what time of the year it was. In response to this critical need, many versions of the calendar were invented. The Anasazi created the Sun Dagger spiral, and the Aztecs created an even more complex calendar known as the Sun Stone. The druids of the British Isles conceived of and erected monolithic structures such as Stonehenge and Newgrange that allowed for the determination of the summer solstice.

One of the most remarkable of the ancient devices for predicting the seasons was based on tracking the movement of celestial bodies such as the moon and the visible planets across the night sky. It was an elegantly conceived and highly advanced instrument constructed around 2200 BP in Greece, made of solid brass and named the Antikythera Mechanism. This remarkable invention, and undoubtedly many more of which we are unaware, were practical instruments, allowing cultures that had adopted some form of permanent agriculture to predict the seasons. Ritual practices evolved from these important landmark seasonal events (planting, harvesting), and most certainly gave rise to many organized religions.

As agriculture became the accepted way of acquiring

food, settled populations grew larger, establishing stable urban communities that, in turn, led to a brilliant series of cultures. Europe, Asia, South Asia, the Middle East, and South and Central America all spawned dominant civilizations that rallied around naturalistic religious concepts revering and celebrating the production of food. As mentioned, the Egyptians worshiped the scarab beetle (Khepri) and the sun (Ra), since they had intuitive knowledge that these two objects plus water were directly responsible for the production of plant life, and thus for all the rest of life, too. The Aztecs and many Asian cultures (e.g., Japan) also revered the sun, presumably for the same reason. They all got it right. We have inherited the legacy of those ancient civilizations and have refined and redefined the ways in which food is provided, as well as reinterpreted the religious tenets upon which those practices were founded.

GO FORTH AND MULTIPLY

During the centuries that followed, farming in many different forms spread to nearly all regions of the world, giving countless numbers of people the advantages of a predictable food supply. With the establishment of cities, trade, and oceanic shipping, the cultivation of plants far removed from their origins became the norm. Invading new soils, crops of wheat, corn, rice, barley, potatoes, and hundreds of others were adopted by burgeoning populations of urban dwellers as their main cuisine items, even though they had no idea where their food was coming from. The only thing that seemed to matter to them was that it was available each year after the harvest. The lack of a good means of preserving fresh produce gave way to prepared foods with long shelf lives. Grain was milled into flour, corn into meal, and rice could be stored as is. Rye seed and other cereals gave rise to storage problems, though, especially in northern Europe. A fungus (*Claviceps purpurea*) routinely infected the stored seed, and when ingested caused an illness, often fatal, that came to be known as Saint Anthony's Fire. Ergot poisoning is what we know it as today. The fact that some harvested crops turned sour did not stop the surge of agricultural initiatives. Nonetheless, many died from food poisoning, mostly caused by bacteria: Salmonella and shigella were as common then as they are today. Fertilizing crops with human feces unwittingly encouraged the spread of these infections, as the germ theory of disease and good sanitary practice did not take hold in Europe until late in the 1800s. Those cultures situated in moderate climates were more fortunate in some ways, enjoying year-round availability of fresh fruits and a variety of vegetables. Storage of these food items was more difficult, however; much of what was harvested had to be eaten that day or the next, or it rotted. Another advantage of warm climates was having multiple harvests in a year of the same crop. In Southeast Asia, for example, three crops of rice were possible.

HOW 'BOUT DEM APPLES

In the meantime, the newly introduced plant species were "morphing" into new kinds of plants we now refer to as cultivars, as farmers selected them for qualities related to taste and overall appearance, rather than for resistance to plant diseases, for instance. Many of these emigrant crops adapted well to shifts in annual temperature profiles and precipitation regimes by growing in ways that differed in surprising ways from the parent plant. A good example is what happened to the apple. Originally from the Tien Shan forests of eastern Kazakstan, this highly sought-after taste treat started out as a small, bitter, pea-size, berrylike fruit. Over time, with our care, it has evolved into some twenty thousand varieties, of which more than seventy-five hundred are raised in commercial quantities. None of these even remotely resembles the original plant product.

Wheat grew almost everywhere, but in Scotland, this wild grass turned keystone crop became adapted to the low light, short growing seasons, and harsh weather patterns of that region. It produced robust, hardy spears laden with the largest, most nutritious kernels of wheat germ in the world. Corn also changed dramatically from its maize ancestral roots, becoming identified with the typical double-ear-bearing tall plant that grows as "high as an elephant's eye"—quite different indeed from its inconspicuous lowly grass plant parents.

PLANTS ON A LEASH

Once we learned how malleable the genetics of plants were after their domestication, vast numbers of new crops were selected from wild plants and tamed in remarkably short periods of time. When they were introduced, many of them changed forever the way people ate. Marco Polo brought pasta back from the Orient, and with the introduction of the tomato from South America into Italy, helped to create a set of cuisines that today is emulated in most of the civilized world. The potato, of which there are some 4,500 varieties today, also originated in Peru and spread into Europe and the British Isles. This starch-laden tuber revolutionized table fare for generations wherever it grew. In fact, there were not too many places it did not grow. It literally thrived in the nutrient-poor soils of Northern Hemisphere climates, making it the ideal addition to many cultures whose usual fare consisted of salted, dried fish. So dependent did some cultures become on "invader" species such as the potato that when these new crops succumbed to plant diseases, eliminating them from the local diet, starvation and even death ensued among large populations; witness the potato famine of the 1800s in Ireland. Mass emigration of Irish people to North America, Australia, and other parts of the world as the result of that agricultural catastrophe redistributed the human gene pool from that small island country, enriching the recipients as well as the donors. Another plant that humans learned to tame very well was rice. Rice cultivation in Asia became almost a religion in itself, as this important grain was established as the keystone crop for the entire region. It remains the basis of most of the cuisines for one-fourth of the world's population.

TROUBLE IN PARADISE

As discussed above, along with all the benefits of farming came the failures. These events were bound to happen no matter what the crop or where it was planted. Adverse weather (floods, droughts), plant diseases, and insect pests all conspired to limit the amount of a given crop. Nature had never planned for monocultures; biodiversity was and still is the rule that enables the establishment of functional ecosystems. Resiliency in nature is related to the number of species a region can support, not the number of individuals of a single species, such as corn or wheat. Granted, many of the grasslands, tundra, and alpine forests harbor just a few dominant species, but there are plenty of other plants and animals interspersed among them to help even out the flow of energy from one trophic (energy) level to the next. Farming excluded any invader that might take away nutrients from the crop of choice. As we will see in the next chapter, we have gone to great lengths to ensure that we get back only that which we plant. It was and still is an unnatural way to behave ecologically. Without being supplemented for depleted nutrients (i.e., fertilized), the soils were not rich enough on their own in most places to support more than a few years' worth of a given plant. Exceptions abound, however: Volcanic ash left over from centuries of geologic activity produced some of the richest soils on Earth, and farming thrived in these regions. And as pointed out earlier, floodplains were also rich in nutrients, as proven by the sustainable cultures of Egypt and Italy.

Yet never did it occur to any human population, regardless of the time period or the fertility of the land, that what they were doing to the environment by farming was actually destroying the very tapestry of what allowed us to evolve into human beings; namely, an intact ecosystem. Instead, we contrived a series of edicts that later became etched in stone that gave us "permission" to lord it over the lowly life forms we could now eliminate from our immediate living space. Therein lies the crux of the problem.

WORLD DOMINATION

A philosophical subtext arose out of most of the popular Western-based religions that were established following the original waves of the first agricultural revolution. It stated unequivocally and without any thought other than to the betterment of humans that God has given us permission to dominate over the land and its natural processes.

The Old Testament was explicit:

> GENESIS 9:1–2. *Be fruitful and multiply, and fill the earth. And the fear of you and the dread of you shall be on every beast of the earth, on every bird of the air, on all that move on the earth, and on all the fish of the sea. They are given into your hand.*

Perhaps the reason all of the earth's wildlife is fearful of us is because we have, indeed, been fruitful and multiplied, and have now filled up the earth to the point of threatening to overwhelm all the rest of the natural world. The strong desire to preside over natural processes has led to a conscious multicultural arrogance that we can in fact do it. When things go our way, we get a false sense of actually being in control of our own destiny. In stark contrast, adverse weather events—floods, cyclones, dust storms, hurricanes, tornadoes, droughts, and heat waves—and other unwelcome natural phenomena, such as tsunamis, earthquakes, volcanic eruptions, and the emergence and reemergence of a wide variety of infectious disease agents, continue to give us pause, and force us to rethink this central theme. Numerous insightful and creative mythologies, and a robust fiction-based literature, pits "man against nature." In the real world, nature usually wins. One colorful bumper sticker sums it up this way: NATURE BATS LAST. In fact, another far less well-known quote from that very same Old Testament reads:

> LEVITICUS 25:23–24. *The land is mine and you are but aliens and my tenants. Throughout the country that you hold as a possession, you must provide for the redemption of the land.*

Apparently, as the story goes, even the Reverend Billy Graham was unaware of this passage when someone read it to him at a debate on Creationism versus evolution.

DON'T WORRY, BE HAPPY

Stewardship is an integral part of our moral contract with the natural world that surrounds us. One place that still exemplifies this concept is the country of Bhutan, a land of gentle, friendly people. I had the privilege of visiting that small kingdom years ago in October, during their harvest season. I was struck by the fact that their religion, a form of tantric Buddhism, did not permit the use of draft animals. All farming was done by hand, right down to the harvesting and winnowing. Their crops were simple but balanced, allowing for a healthy life: rice, wheat, chili peppers, tomatoes, and some leafy green vegetables. The entire country was fully engaged in the harvest when my wife and I arrived. Many of the festivals we attended celebrated this singular event. On one memorable occasion, we witnessed six people on a hillside lined up in two rows on either side of a blue plastic tarpaulin, their only apparent concession to anything modern. Each held a wooden two-pronged pitchfork. These hand-crafted farm implements were indistinguishable from those used by the world's first wheat farmers some ten thousand years ago. One person stood at the end of the tarp and tossed a healthy bunch of newly cut buckwheat into the air. The other five, two on one side and three on the other, immediately beat the wheat down onto the tarp, knocking the grains of wheat off the spears. The unseparated mixture was then taken to a field lower down in the valley where the wind blew strongly in one direction. Later that day, in that same valley, my wife and I watched awestruck as one woman stood with her back to the wind and winnowed the grain from the chaff by pouring large baskets of the beaten mixture downwind. The weight of the grain allowed it to fall at the feet of the winnower, while the lighter chaff accumulated some ten feet away. Again, I could not help thinking that I had time-traveled back to the very origins of agriculture. These are images I will take to my grave.

TROUBLE IN PARADISE REDUX

Bhutan has a population of around seven hundred thousand. Because almost everyone farms, they produce more than enough food for themselves, with enough left over to export to places like India in exchange for a few of the modern essentials, such as gasoline. Bhutan's population will never exceed its food supply as long as everyone takes part in the process of producing it. But even in this idyllic society, there are problems looming on the horizon. Bhutan's current biggest problems are related to urbanization; obesity, heart disease, and illiteracy seem to be foremost on the mind of the minister of health. The hope is that they will reach a balance with their conscious intent to modernize. It would be a shame to see this example of a self-contained society collapse for the same reasons as so many others have done in the past.

A MODERN SYNTHESIS

The current collective worldview of how we should conduct our lives in context with the rest of the earth's living entities recognizes the same basic facts, regardless of the culture: namely, that nature is never wholly predictable, that it often poses threats to our very existence, and, above all, that it can never be fully understood. The development of the modern applied science of public health has added hard data in support of that notion, and has led to the following realization regarding the consequences of altering the terrain for whatever purpose, be it agriculture, settlement, or industrial development: We are at risk of acquiring an illness (albeit unpredictable and certainly unintended) related to any human activity that significantly rearranges the natural landscape. The global scientific community is rapidly coming to consensus that the way in which we must carry out our lives at both the individual and population level, and at the same time avoid adverse health consequences, is to strive to achieve a degree of ecological balance with the rest of the earth's life forms.

CHAPTER 3

TODAY'S AGRICULTURE

There is nothing like returning to a place
that remains unchanged to find
the ways in which you yourself have altered.

—NELSON MANDELA

SOUR GRAPES

To get a feel for what it must have been like to live at the pinnacle (i.e., the last gasps) of the first agricultural revolution, we need look no further than to John Steinbeck's take on farm life, set in what can only be described as one of the all-time worst periods of American history. His classic novel *The Grapes of Wrath* was published in 1939, but he wrote most of it during the height of the Great Depression, while all the tragedies of that period were unfolding right in front of him. Steinbeck focused on the plight of the farmer and agriculture, in general. He used the fictional trials and tribulations of a typical rural American family from Oklahoma after their farm failed, as so many other farms had also done, to illustrate what U.S. farm life had become. In simple language and with powerful declarative sentences, the author puts the reader smack on the back of the Joad family truck as they slowly wend their way westward toward the land of "milk and honey," the Central Valley of California. They had been "evicted" from their homestead by one of the most severe and long-lasting droughts in that region since records were kept. It is important to note that droughts are part of the normal precipitation pattern for all tall grass prairies. Viable farming operations for most domesticated crops are impossible in those semiarid ecosystems, as wheat and barley cultivars require significant quantities of water. But farm they did. Homesteading in what was to become known as the "dust bowl" even had the official seal of approval from the U.S. government. But after about twenty unusually wet years (1910–1930), the whole thing went sour. Apparently, nothing could undo what these well-intentioned dirt farmers had inadvertently done; that is, to lay bare the landscape by plowing and planting. Not even the most sincerely offered prayers had any effect on reversing the disaster they now faced.

So without any options left to the Joads but to abandon their land, into their trustworthy Model T Ford truck went virtually everything they owned, including the pets. Farm animals had no luck at all and were by necessity left behind to die a slow, painful death by starvation, dehydration,

[PROBLEM]

Feeding the World: Another Brazil

Growing food and raising livestock for 6.8 billion people require land equal in size to South America. By 2050 another Brazil's worth of area will be needed, using traditional farming; that much arable land does not exist.

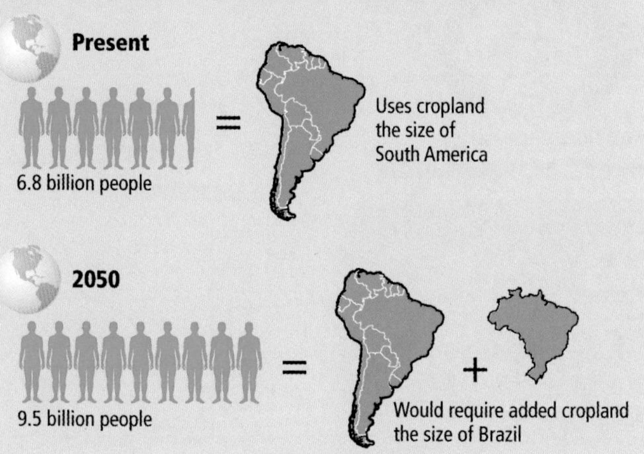

Courtesy Laurie Grace

AGRICULTURE AND LAND USE Fifteen thousand years ago, there were no farms on Earth. Today, we raise our crops on a landmass equivalent to the size of South America. If we throw animal husbandry into the mix, we use some eighty percent of all the available dry land. Population experts predict that within another forty to fifty years, there will be another three billion of us. That is a lot of new mouths to feed. If we continue to farm in the traditional fashion (i.e., soil-based), then we would need another Brazil's worth of land to farm to produce crops. This much new arable land does not exist. Another solution is required if we are to avoid massive starvation and armed conflicts caused by the scarcity of essential resources like food and water.

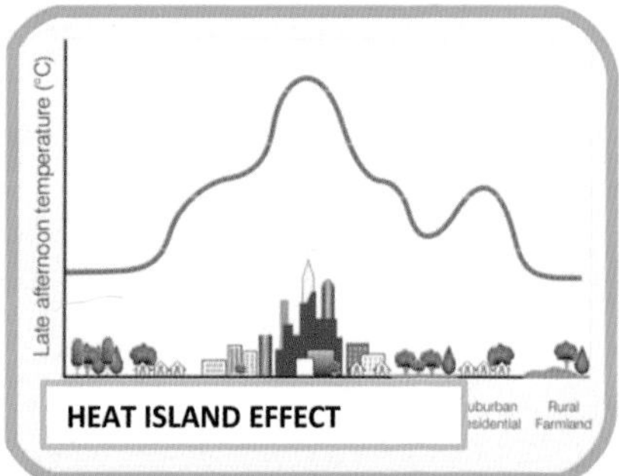

Courtesy Oliver Foster / O Design (www.odesign.com.au)

(*left and opposite*) ADVANTAGES Vertical farms can offer the perfect solution to the main crises the world faces today, such as deforestation, population increases, climate change, pollution, depleting resources, dwindling ecology, decreasing food supplies, urban heat, island effect, and more.

THE VICTORY GARDEN Propaganda poster, produced by the United States government during World War II, encouraged every American citizen to help win the war by growing their own vegetables in "victory gardens," thereby taking some of the pressure off commercial farmers, whose responsibility it was to supply sufficient quantities of staples for the troops overseas. Apparently this campaign worked like a charm, as millions of novice farmers began planting such edibles as cucumbers, corn, and watermelon in their backyards. Local produce soon became the norm, and consumers grew more aware of the value of a freshly picked tomato. After the war ended in 1945, America went back to "business as usual," preferring to rely on the mass production of crops to feed their families. Today, there is a movement to return to locally produced crops, only this time its epicenter is situated within the urban landscape, not in the suburban backyard.

WAR GARDENS FOR VICTORY

GROW VITAMINS AT YOUR KITCHEN DOOR

Enter VICTORY GARDEN CONTEST

REGISTER 404 S. 8th ST. MAY 1-15

Courtesy Minnesota Historical Society

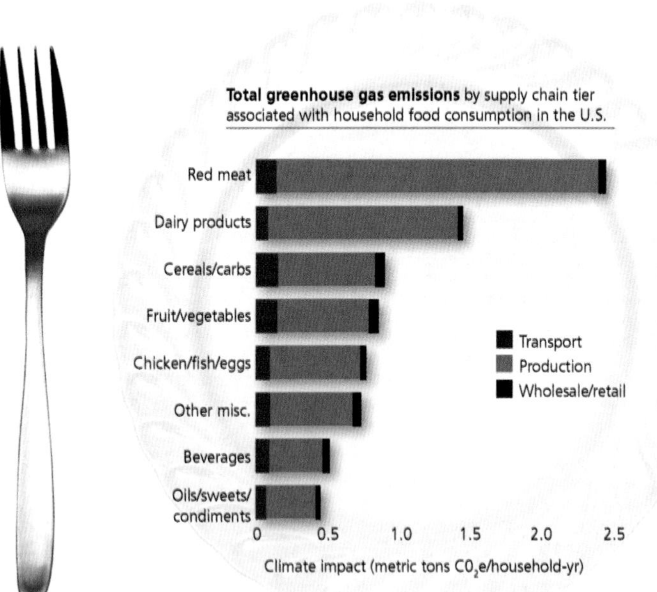

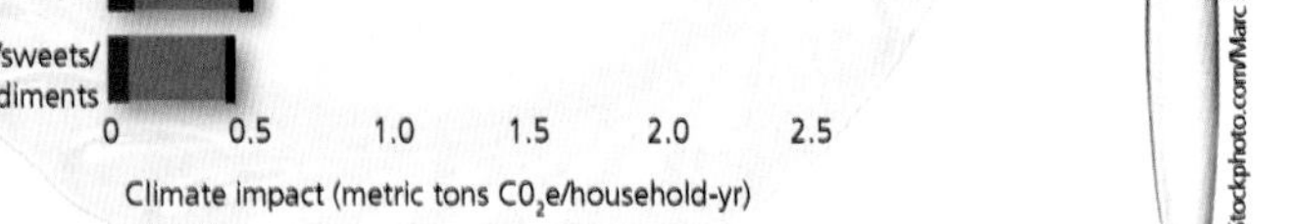

Courtesy Conservation Magazine

© iStockphoto.com/Marc Dietrich

ENDANGERED COASTAL AREAS

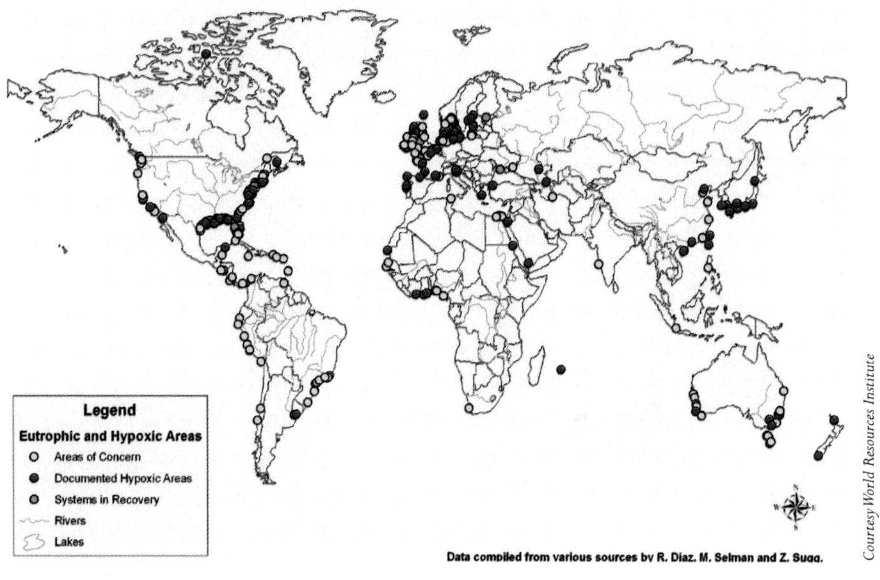

Courtesy World Resources Institute

Agricultural runoff is the world's most destructive source of pollution. Estuaries occur where rivers enter the ocean. Around the world, just within the last twenty years, these highly productive aquatic environments have been overwhelmed by millions of tons of runoff laden with silt, pesticides, herbicides, and nitrogen fertilizers. The result has been a near sterilization of many of the world's most important estuaries.

Nitrogen fertilizers (ammonium nitrate and urea) have been solely responsible for killing off hundreds of billions of immature crustaceans, mollusks, and fish. The mechanism is simple: Nitrogen severely depletes oxygen levels and creates a situation in which larval life forms suffocate and die. It is the prime reason why the United States has to import over eighty percent of its seafood every year.

The flooding of farmland along major rivers is due, in part, to a rapidly changing climate, along with the overuse of agrochemicals. Eutrophic areas happen where the ocean has been overloaded with nutrients. Hypoxic zones are places with low oxygen content in the water.

Courtesy Getty Images

THE DUST BOWL This haunting photograph shows the dust bowl, caused by the inappropriate application of agriculture in the central Midwestern portion of the United States. Farming on grassland was never a good idea, because the annual precipitation and temperature profiles of that ecosystem do not favor the growth of domesticated crops such as wheat and corn without the application of irrigation strategies and the use of fertilizers. Without these two "forcing" methods, the result was highly predictable; depletion of nutrients from the soil and loss of crops due to insufficient rainfall. Ultimately, the soil failed, farms went under, and the farmers moved on to parts West, mainly California.

(*opposite*) CALIFORNIA'S CENTRAL VALLEY Over the last fifty years, California has mobilized water resources from as far away as Colorado. Irrigation projects of all sizes and capacities crisscross the Central Valley and supply much-needed water for a wide variety of crops, including table grapes, almonds, oranges, and avocados. Most of the irrigation methods are of the flood variety, in which water is allowed to flow over the crops in a lakelike fashion several inches deep, as opposed to flowing through irrigation ditches.

The application of fertilizers, herbicides, and pesticides insures that each crop grows to its maximum yield. Unused portions of these agrochemicals seep into the ground,

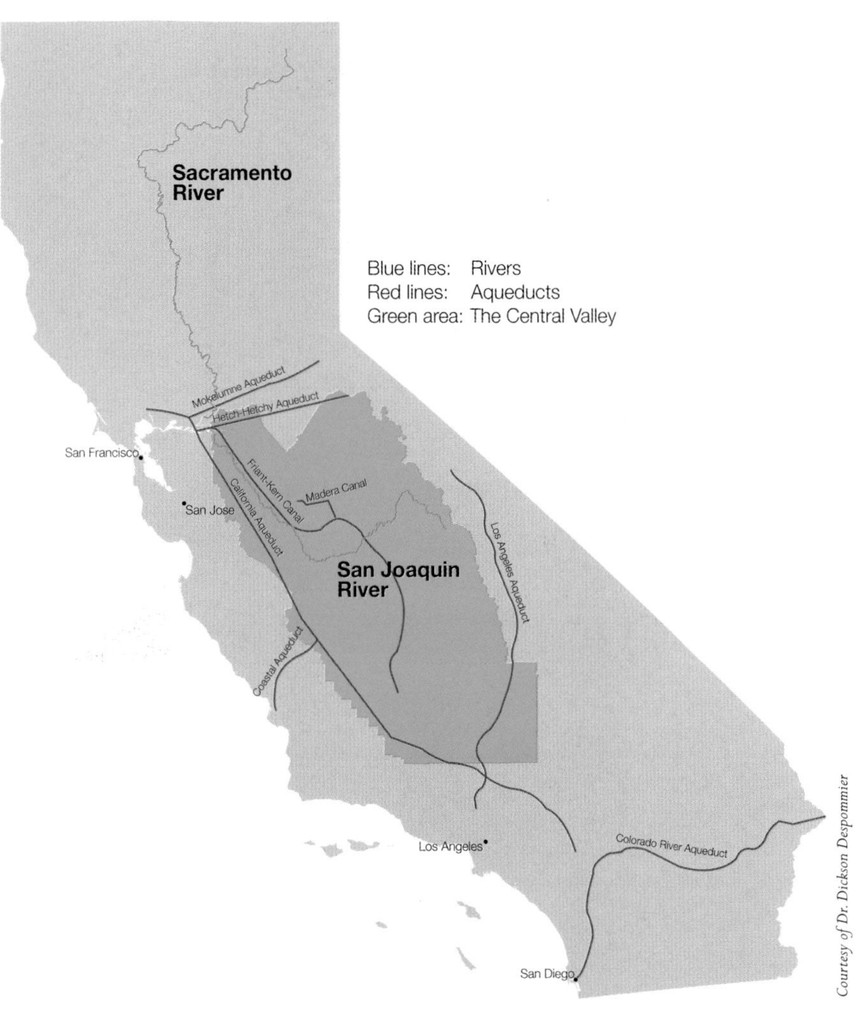

Courtesy of Dr. Dickson Despommier

contaminating ground water. In the southern Central Valley, this has resulted in the accumulation of salt-laden water in the aquifer. Because there is no place for the water to go (no rivers or outlets to the ocean), it continues to rise up towards the deepest taproots. When it reaches them, those plants will surely die.

Some predict that in another twenty five years California will no longer be able to farm in that vast southern region, resulting in an estimated loss of more than $30 billion in agricultural revenue. In the north, agricultural runoff eventually ends up in the Sacramento River, and ultimately into San Francisco Bay, and creates yet another set of ecological disruptions that will ultimately affect all those living in the area.

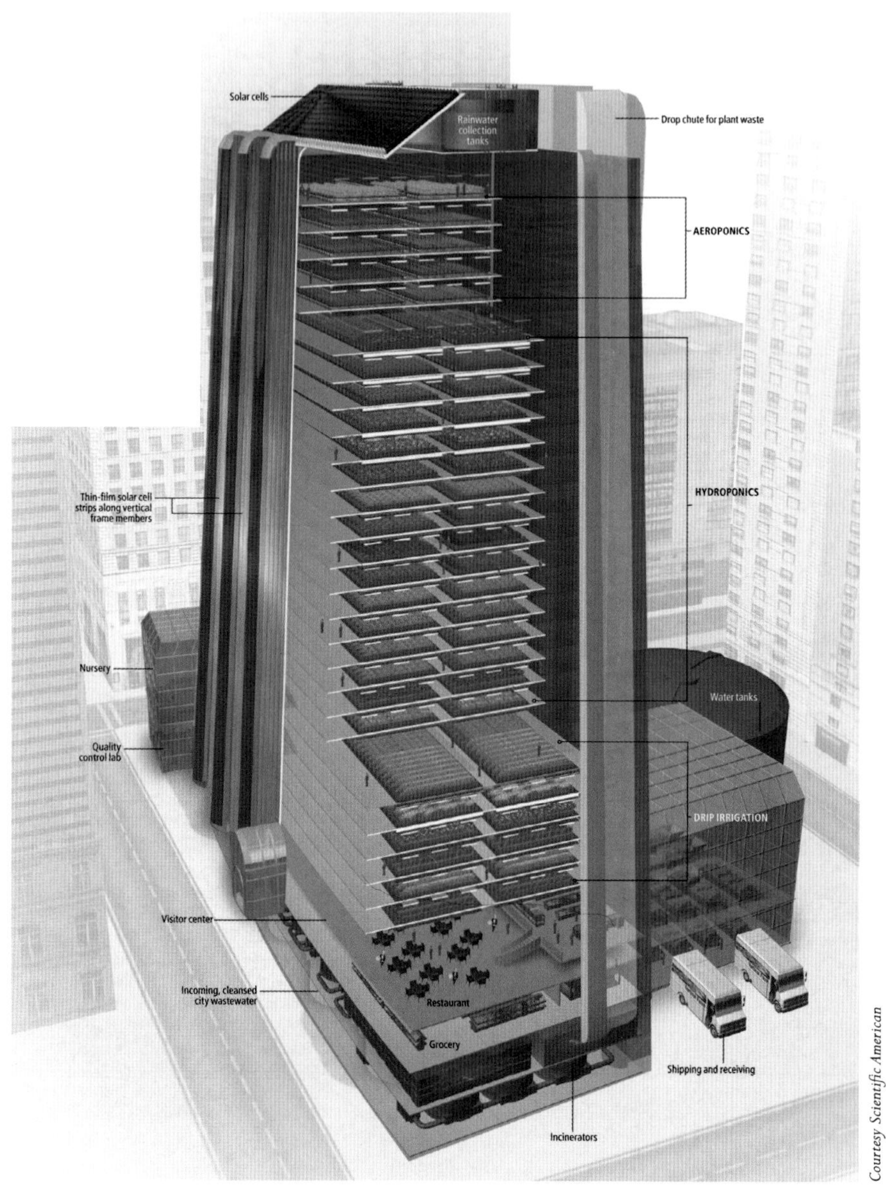

Courtesy Scientific American

HIGH-RISE CROPS A thirty-story vertical farm would exploit different growing techniques on various floors. Solar cells and incineration of plant waste dropped from each floor would create power. Cleansed city wastewater would irrigate plants instead of being dumped into the environment. The sun and artificial illumination would provide light. Incoming seeds would be tested in a lab and germinated in a nursery. And a grocery and restaurant would sell fresh food directly to the public.

Courtesy of Dieter Tamson

HYDROSTACKERS The hydrostacker is a cleverly designed device for saving acres of floor space and still allowing the optimal growth for a wide variety of hydroponically grown crops, including strawberries. The hydrostacker has revolutionized the greenhouse industry, and at the same time, allowed for the return of huge tracts of land to natural process. In one instance, a Florida farmer, wiped out by hurricane Andrew, reinvested in a greenhouse and replaced some thirty acres of outdoor farmland with a single acre of greenhouse-grown strawberries using hydrostackers. Ideas like this can radically change the way we think about agriculture and land use in favor of ecosystem restoration.

Courtesy of Jung Min Nam / JN_Studio, Thesis Project at GSD Harvard University, 2009 / Advisor: Prof. Ingeborg Rocker

(*above and right*) HYDROPONICS GROWING SYSTEM WITH NUTRIENT FILM TECHNIQUE (N.F.T) SYSTEM Hydroponics is a system where plants are grown in nutrient water that has been enriched with minerals, and is used to optimize yields and quality of produce, especially for the indoor space. Hydroponics is a technically sophisticated and well-established commercial practice in most regions of the world. It can be the best available technology for growing vegetables in cities—especially within a building—and offers a high yield with a relatively small footprint.

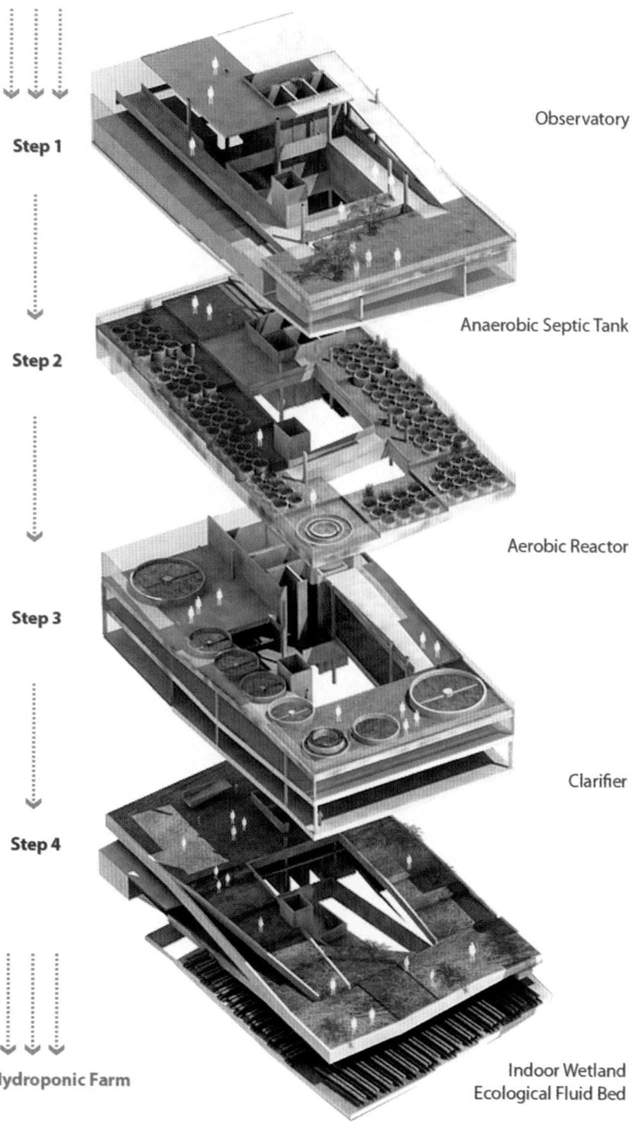

Courtesy of Jung Min Nam / JN_Studio, Thesis Project at GSD Harvard University, 2009/ Advisor: Prof. Ingeborg Rocker

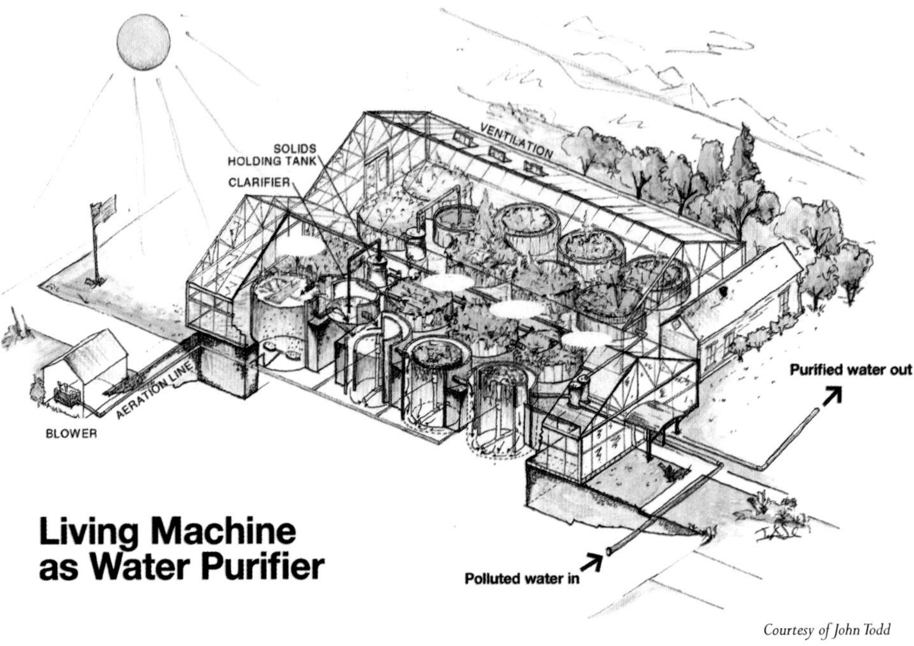

Courtesy of John Todd

(*above and right*) SCHEMATIC FOR A "LIVING MACHINE" FOR REMEDIATION OF BLACKWATER John Todd is credited with coining the term living machine for plants that help us remediate damaged aquatic ecosystems. He worked diligently in the 1960s to identify plant species that would remove toxic materials (heavy metals, pesticides, herbicides, fertilizers) from damaged lakes, wetlands, and estuaries. Many of his original findings have been successfully applied on a commercial level and are still in use today. One shining example is located in White River Junction, Vermont, and employs plants to return the blackwater (human feces and urine) into usable water for reuse in toilets. The system has performed admirably and is a prime example of the simplicity and efficiency of enlisting the help of plants to permit us to live more in sync with nature.

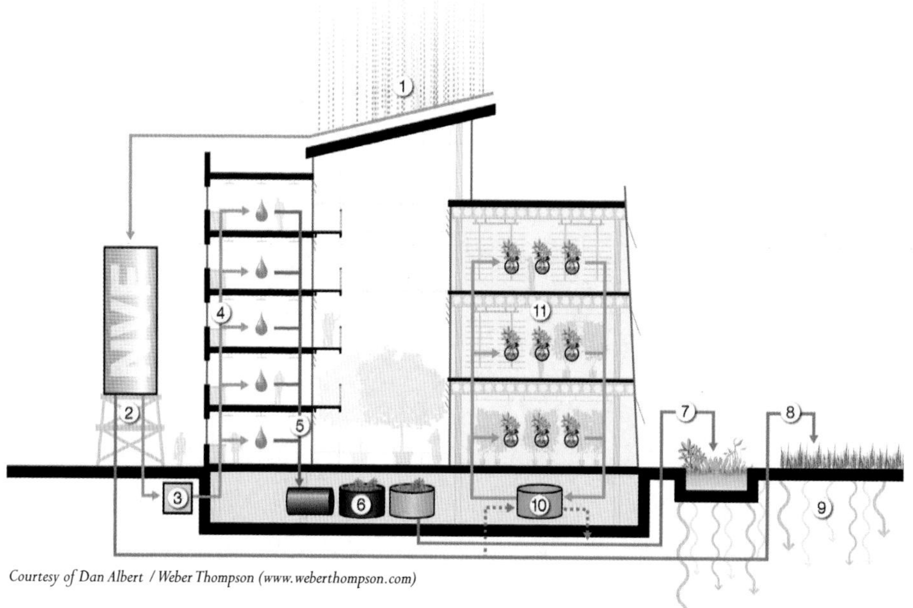

Courtesy of Dan Albert / Weber Thompson (www.weberthompson.com)

WATER SYSTEM

1. rain water collection
2. cistern
3. purification
4. potable water
5. grey/black water
6. on-site wastematter treatment
7. output water to wetland system
8. rain water for urban farm
9. on-site infiltration
10. nutrient supply for growing systems
11. hydroponic, aeroponic growing facility

Courtesy of Rachel Hiles

HYDROPONIC GROW SYSTEMS A typical hydroponic greenhouse uses low-cost polyvinyl chloride (PVC) plastic piping to hold plants in place for the delivery of plant nutrition.* Nutrient film technology is the formal term applied to modern hydroponic growing systems, in which a thin stream of nutrient-laden water is run over the root system of crops. Hydroponics uses some seventy percent less water when compared with conventional irrigation schemes employed in outdoor soil-based agriculture. What is more, hydroponic growers do not produce any agricultural runoff. Indoor farming allows for the complete recovery of the water of transpiration by dehumidification. Because of these advantages, hydroponic farming is applicable to any situation in which there is sufficient water and sunlight, regardless of outside soil type or weather patterns. The hydroponics industry is rapidly growing both in the number of successful operations and in the expansion of the variety of plants that can be grown commercially indoors.

* See page 282.

Courtesy of Richard Stoner / AgriHouse (www.agrihouse.com)

THE AEROPONIC SYSTEM Aeroponics is the application of a fine mist of water laden with plant nutrients onto the root system of a given crop, in this instance, spinach. The roots are enclosed in a chamber that keeps the humidity at a maximum level. Aeroponics was invented by Richard Stoner while working for NASA. This technique uses seventy percent less water than hydroponics, making it a highly desirable method of indoor farming, particularly in regions where water is at a premium. Virtually any plant can be grown in this fashion.

THE PERIODIC TABLE OF THE ELEMENTS

1 **H** Hydrogen 1.00794																	2 **He** Helium 4.003
3 **Li** Lithium 6.941	4 **Be** Beryllium 9.012182											5 **B** Boron 10.811	6 **C** Carbon 12.0107	7 **N** Nitrogen 14.00674	8 **O** Oxygen 15.9994	9 **F** Fluorine 18.9984032	10 **Ne** Neon 20.1797
11 **Na** Sodium 22.989770	12 **Mg** Magnesium 24.3050											13 **Al** Aluminum 26.981538	14 **Si** Silicon 28.0855	15 **P** Phosphorus 30.973761	16 **S** Sulfur 32.066	17 **Cl** Chlorine 35.4527	18 **Ar** Argon 39.948
19 **K** Potassium 39.0983	20 **Ca** Calcium 40.078	21 **Sc** Scandium 44.955910	22 **Ti** Titanium 47.867	23 **V** Vanadium 50.9415	24 **Cr** Chromium 51.9961	25 **Mn** Manganese 54.938049	26 **Fe** Iron 55.845	27 **Co** Cobalt 58.933200	28 **Ni** Nickel 58.6934	29 **Cu** Copper 63.546	30 **Zn** Zinc 65.39	31 **Ga** Gallium 69.723	32 **Ge** Germanium 72.61	33 **As** Arsenic 74.92160	34 **Se** Selenium 78.96	35 **Br** Bromine 79.904	36 **Kr** Krypton 83.80
37 **Rb** Rubidium 85.4678	38 **Sr** Strontium 87.62	39 **Y** Yttrium 88.90585	40 **Zr** Zirconium 91.224	41 **Nb** Niobium 92.90638	42 **Mo** Molybdenum 95.94	43 **Tc** Technetium (98)	44 **Ru** Ruthenium 101.07	45 **Rh** Rhodium 102.90550	46 **Pd** Palladium 106.42	47 **Ag** Silver 107.8682	48 **Cd** Cadmium 112.411	49 **In** Indium 114.818	50 **Sn** Tin 118.710	51 **Sb** Antimony 121.760	52 **Te** Tellurium 127.60	53 **I** Iodine 126.90447	54 **Xe** Xenon 131.29
55 **Cs** Cesium 132.90545	56 **Ba** Barium 137.327	57 **La** Lanthanum 138.9055	72 **Hf** Hafnium 178.49	73 **Ta** Tantalum 180.9479	74 **W** Tungsten 183.84	75 **Re** Rhenium 186.207	76 **Os** Osmium 190.23	77 **Ir** Iridium 192.217	78 **Pt** Platinum 195.078	79 **Au** Gold 196.96655	80 **Hg** Mercury 200.59	81 **Tl** Thallium 204.3833	82 **Pb** Lead 207.2	83 **Bi** Bismuth 208.98038	84 **Po** Polonium (209)	85 **At** Astatine (210)	86 **Rn** Radon (222)
87 **Fr** Francium (223)	88 **Ra** Radium (226)	89 **Ac** Actinium (227)	104 **Rf** Rutherfordium (261)	105 **Db** Dubnium (262)	106 **Sg** Seaborgium (263)	107 **Bh** Bohrium (262)	108 **Hs** Hassium (265)	109 **Mt** Meitnerium (266)	110 (269)	111 (272)	112 (277)	113	114				

58 **Ce** Cerium 140.116	59 **Pr** Praseodymium 140.90765	60 **Nd** Neodymium 144.24	61 **Pm** Promethium (145)	62 **Sm** Samarium 150.36	63 **Eu** Europium 151.964	64 **Gd** Gadolinium 157.25	65 **Tb** Terbium 158.92534	66 **Dy** Dysprosium 162.50	67 **Ho** Holmium 164.93032	68 **Er** Erbium 167.26	69 **Tm** Thulium 168.93421	70 **Yb** Ytterbium 173.04	71 **Lu** Lutetium 174.967
90 **Th** Thorium 232.0381	91 **Pa** Protactinium 231.03588	92 **U** Uranium 238.0289	93 **Np** Neptunium (237)	94 **Pu** Plutonium (244)	95 **Am** Americium (243)	96 **Cm** Curium (247)	97 **Bk** Berkelium (247)	98 **Cf** Californium (251)	99 **Es** Einsteinium (252)	100 **Fm** Fermium (257)	101 **Md** Mendelevium (258)	102 **No** Nobelium (259)	103 **Lr** Lawrencium (262)

Courtesy of Rachel Hiles

Orange indicates elements required by humans only.

Green indicates elements required by both plants and animals, including humans.

or both; millions of cattle, sheep, pigs, and chickens were left scattered around the four states most affected by the drought. On the road, the Joad family suffered one disaster after another, but they persevered and finally settled in California, only to encounter a whole new set of social problems. Steinbeck was indeed a savvy observer of the human condition as a whole generation of Americans felt the effects of the damage they'd done to the natural environment. Soon that same generation would enter the bloody conflict of World War II.

The Joads are a skillfully crafted and highly measured collection of American stereotypes, all of whom resonate well with the "common man" theme for which Steinbeck was so famous. He was their bulldog, their champion, their chronicler. He wrote of the injustices confronting most people living in what we would still classify today as an advantaged, enlightened, developed part of the world. Yet *The Grapes of Wrath* remains one of the most scathing indictments of man's inhumanity toward man, a darkly painted canvas of grief showing the raw underbelly of government's and management's total lack of concern for the welfare of destitute America during the Depression years. The only thing that even comes close in recent U.S. history is the Bush administration's handling of the mess that followed after Hurricane Katrina struck. Steinbeck would surely have had a field day with that political disaster.

Steinbeck depicts farm life in its worst-case scenario. Management versus unionism, poor dirt farmers trying to eke out a living in a region of the country that was never environmentally suited for crop production, at least not without massive irrigation projects. So powerful and truthful was his writing that it literally changed U.S. and world history. For it, he was awarded the Nobel Prize in Literature and the Pulitzer Prize for best novel. I am continually haunted with each rereading of this depressing story by images and refrains from the collective genius of Woody Guthrie, Pete Seeger, and Ramblin' Jack Elliot, whose pro-union songs echo down the rutted back roads of the desertified Midwest; ghost convoys of overloaded, nearly worn-out vehicles of all sizes and shapes shrouded in a dirt-laden cloud of former farmland, a silence of mass exodus, and of the destitute families who elected to remain and eventually, when all hope was gone, committed suicide in their filthy, dilapidated hovels; the half-buried skeletons of legions of farm animals littering the barren countryside like some collaborative surrealistic hybrid painting by Georgia O'Keeffe and Francis Bacon.

In 535 pages, *The Grapes of Wrath* brings to a close the final chapter of an agricultural adventure that lasted some ten thousand years. Not a bad life span for any revolution, let alone one involving the actual reshaping of the very environment that spawned us. I once showed the Academy

Award–winning film version to a class of mine and was shocked to learn that most of them had never even heard of the novel, let alone the film. Even more shocking was their admission that they did not know who Henry Fonda was. Hoping to somehow connect with these science-driven youth in an otherwise palpable cultural generation-gap moment, and in an obviously heightened state of frustration, I yelled: "For God's sake, he was Bridget Fonda's grandfather!" They all relaxed and smiled: "Oh, that Fonda!" There wasn't a dry eye in the class at the end of its showing, including mine. Many of the students then went on to read the book and were deeply moved by its message.

FOUR PLAY

The origins of twenty-first-century agriculture can be traced back to the convergence of four things: the American Civil War, the discovery of oil, the development of the internal-combustion engine, and the invention of dynamite.

The War Between the States nearly consumed a nation divided. Of some 33 million U.S. citizens alive in 1860, nearly 4 million died over the next four-year period, starting on April 12, 1861, and ending on April 9, 1865. Many were civilians. So what else is new? It was the birth of the sharp-shooters. Walt Whitman wrote passionately about the war, and young Winslow Homer set off from New York City to illustrate it for *Harper's Weekly.* The Civil War was a fight to the finish over the right to control what happened to the country's cotton crop, the South's main agricultural product. Opposing slavery was almost an afterthought. New England–based textile and clothing manufacturers wanted unlimited access to the gin-milled cotton and wanted it on their own terms, while the Southern cotton growers wanted the best price for their harvest, no matter who bought it. Conflict ensued, big time. The South ended up selling most of its cotton production to Europe at substantially higher prices than the Northern industrialists could or would agree to pay. In either case, the Yankees reacted quite negatively to Dixie entrepreneurship. For the Southern plantation owners, labor was dirt cheap, enabling them to achieve outrageous returns on their harvests. That's because they relied on slaves brought over mainly from West Africa as their primary labor force, although indentured white farmworkers were plentiful, too. Besides providing a reliable work force, slaves from West Africa did one other thing that would eventually bring about another remarkable change to the New World: They introduced hookworm parasites into American soil.

As the savagely fought conflict progressed year after year, army recruiters began to feel an increasing lack of support throughout most of the North. A rallying slogan was what was called for.

"Help Bring Back the Cotton to Our Mills So a Few of Us Can Get Filthy Rich"

This "battle cry" would not have resonated all that well with young recruits, nor with anyone else for that matter. Something else was required to differentiate the North from the South. Something with an emotional and moral kick to it. In fact, something that would allow the Northern protestant ethic its fullest range of expression. They settled on the issue of slavery. Abolitionists had repeatedly petitioned Congress to outlaw it long before the war had even started. Now their idea was gaining traction, but for all the wrong reasons; cynics have had a good time pointing out all the false morality issues surrounding the final decision to go ahead with the abolition proceedings. The anti-slavery movement got its way a full two and a half years after the first shots at Fort Sumter were fired, when it was apparent to all, both North and South, that the North might actually lose thanks to General George B. McClellan's exceptional aptitude for screwing up even the simplest of battle plans. On January 1, 1863, President Abraham Lincoln finally issued his famous Emancipation Proclamation making slavery illegal. Without slaves, Southern gentlemen would be forced to do all the hard labor themselves. This was never an option, so the war continued on for yet another two and a half years. The turning point occurred in 1864, with the replacement of McClellan with Ulysses S. Grant. The pivotal battle of Gettysburg was won by General George Meade of the Army of the Potomac, who defeated General Robert E. Lee; shortly thereafter, the South surrendered at Appomattox Court House. The war had ended. The rest of the world rejoiced, and slavery was indeed abolished, at least in name. The South sank into an economic blue funk that lasted for some twenty years thereafter, and the Northern mills ended up getting their cotton from places such as India, Central America, and Egypt.

If the South was ever to rise again as an agricultural force to be dealt with (and rise it did), then a new way of farming had to be invented. Enter the discovery of oil and the development of the internal-combustion engine. A political battle for economic supremacy became the singular event that eventually forced Southern landowners to switch to mechanized farming equipment, replacing human labor with gas-guzzling clunkers. It would be the machines, and the many agricultural innovations that took advantage of them, that helped define the second great agricultural revolution.

New Oil

The first site for the production of significant quantities of crude oil in America was Titusville, Pennsylvania. In 1859 "Colonel" Edwin Drake drilled not so deep into the oil-seeped bedrock and struck pay dirt. Oil had been discovered even earlier in Poland, in 1854. This sticky black liquid launched an industry that rapidly spread all over the world, even into Texas. Today the OPEC countries of the Middle East are legend, and oil and natural gas stand alone as the world's two most important concentrated sources of energy for nearly everything, including the operation of complex farming equipment. For reasons that need not be spelled out since they are so widely accepted, the burning of oil products has also become the bane of planet Earth's existence.

Internal Combustion

Nonetheless, the discovery of oil was not fully appreciated by all of the inventors of the internal-combustion engine. It was Nikolaus August Otto in Germany in 1861 who discovered that by compressing just the right mixture of air and gasoline in a confined space, then igniting it, enough energy could be released to drive a piston that, in turn, made a flywheel go around, producing work. It was that simple. It should come as no surprise that the first car manufacturers to take advantage of this finding were also located in Germany. Prior to that, steam was used to propel cars, but a number of technical difficulties—including boiler explosions and meltdowns—made traveling somewhat unpredictable.

The Stanley Steamer and all its relatives were doomed from the start, as oil became the fuel of choice, even though the Steamer never had a boiler failure that hurt anyone. Then along came Henry Ford, whose innovative ideas resulted in the creation of the assembly line, standardized parts, and a cheap-to-manufacture affordable car that could run on either gasoline or ethanol. This might have created quite a flurry of agricultural activity toward establishing crops that ethanol could easily be made from (corn and other grains) if the politically righteous right had picked, instead of alcoholism, some other nasty habit to focus on, like smoking. But instead, they pressed for Prohibition and most alcohol production came to a dead stop in 1920. Smugglers, gangsters, and moonshiners rejoiced in unison, perhaps even raising a glass or two of their favorite beverage of choice. Some conspiracy enthusiasts feel it was the oil industry that, behind closed doors, put the quash on ethanol fuels by financially supporting the passage of the Eighteenth Amendment and the Volstead Act. Together, these two pieces of legislation made it illegal to manufacture or sell liquors with high alcohol content. In defense of a nonconspiracy explanation, though, on a worldwide basis, gasoline was indeed the choice of all the early manufacturers of automobiles. It's hard to imagine that the embryonic oil cartel could have engineered such a global solution; it was probably the abundance of crude oil, the ease of making gasoline from it, and its efficiency of burning that made oil the preferred fuel instead.

Henry Ford is also credited with inventing the diesel-powered tractor in 1907. Its introduction rapidly replaced the clumsy, excessively heavy steam-powered tractors popular during the 1800s that often got stuck during springtime planting efforts, especially in already soggy bottomland. They routinely had to be hauled out by teams of horses. Ford's reasonably priced, lightweight, small, agile farm vehicles rarely got stuck, and the moment they hit the plowing fields they completely revolutionized the way agriculture was practiced, although they did not become widely used until the outbreak of World War I. Today, most of the tractors and other farming equipment in the United States are manufactured by the John Deere Company, headquartered in Moline, Illinois, but hundreds of companies make them worldwide. All use gasoline as their fuel of choice. It's no wonder, then, that farming consumes some 20 percent of all fossil fuel used in the United States.

Life's a Blast

In 1847 Ascanio Sobrero, working in his laboratory in Turin, Italy, synthesized the first batch of nitroglycerine, a highly unstable compound, and enabled countless farmers around the world (and a few safecrackers, too), to blow up just about anything they wanted. Unfortunately, it also blew up more than a few of those who used it. In fact, Alfred Nobel's own brother was killed in an unanticipated explosion in the family nitroglycerine factory in Stockholm, Sweden. Undaunted, and under some pressure from local officials, Alfred moved the whole operation to the outskirts of his native city and continued to tinker with the most explosive substance known to humankind up to that point. Between 1864 and 1867, Nobel discovered that mixing nitroglycerine with clay, making a sort of slurry out of it, stabilized the molecule, rendering it harmless no matter the circumstances. It could be dropped, kicked, even stomped on without so much as the slightest bit of reaction. He dubbed the new product "dynamite." Today, we mix nitro with common sawdust to produce the same stable mix. Wrapped into thick paper-covered foot-long sticks with a primer fuse and a percussion cap, dynamite could be safely shipped anywhere in the world. It rapidly became the explosive of choice for clearing land. Stumps that once required teams of draft animals and days of effort to pry them out of their rooted strongholds could now be removed from the forested landscape in less than a single day's work. Empty fields, once virgin woodland, were transformed into domesticated agricultural land. Plowed, then planted with crops like corn and sorghum, virtually any crop could bring in a profit in the early days of Midwest farming. The forest floor was rich in deep, black soil, an ideal situation for any crop species with a penchant for growing in a temperate zone.

DOWN WITH TREES!

In the early colonial days, after clear-cutting the forests to make wood available for houses and fuel, New Englanders tried their hand at farming, to no avail. This despite the fact that the friendlier Native Americans helped many a colonist get started with corn. A six-inch-deep hole in the ground, a small fish, and a kernel of corn was all it took, but Europeans—unfamiliar with the need to fertilize in their new homeland—often left out the fish part, and many got the whole thing wrong from the start. Crop failure was essentially a death sentence. Many perished in the early days of settlement due to the lack of a reliable food supply. The soils were too thin and rocky, and even after years of settlers clearing enough land to plant a few staple crops, these farms often failed due to the short summers and long, freezing winters. Dairy cattle and milk products soon replaced farming in the rock-strewn fields of Massachusetts, Rhode Island, and New Hampshire. In Vermont cheese making and maple syrup production made up for some of the revenue lost due to the impossible farming conditions found there.

As farming spread to other parts of the new colonies, New England became better known as a supplier of hardwood furniture and a center for cloth and leather manufacturing—with water-powered mills of all kinds at the heart of it all—than as a producer of food.

The Northeast rapidly grew back the trees cut down for farming. This was because in 1775, Daniel Boone, along with a hardy bunch of like-minded adventurers, breeched the Cumberland Gap, paving the way for a new cohort of immigrants from Europe. It was to become the gateway to the upper Midwest and its verdant, fertile valleys and floodplains. The rivers that wound sinuously just beyond the Appalachian Mountains and whose headwaters lay in the upper reaches of the western slopes of that same ancient terrain—the Susquehanna, Allegheny, and Monongahela—helped to shape the landscape. Together with the Ohio and Tennessee rivers, these waterways became the commercial arteries to Ole Man River, the mighty Mississippi, making this region of North America an ideal place to settle down, farm, and then transport produce by boat to New Orleans and out to world markets. News of the promised land to the west spread fast, and a horde of new wannabe farmers crossed through the Gap and into the history books. The land was rapidly cleared of its hardwood forests, and farming was soon on its way to becoming the most popular work activity in America. By the time Steinbeck began writing about the demise of farms, one in four people lived on one. Food was soon arriving from the Midwest to all parts of the world. New England settled back into an almost exclusively industrial mode, using the leather from slaughtered dairy cows for shoe manufacturing and continuing with its rich tradition of furniture making. In the mid-1800s, the far West was settled and cattle ranching took over as the number-one provider of a nationwide stable food supply, and created a lucrative by-product, leather, to boot.

The Louisiana Purchase was made in 1803, and an explosion of exploration ensued. America flexed its new muscle in the War of 1812, and during the time between these two events, even more new farmland opened up as wagon trains by the thousands headed out of Saint Louis—Westward Ho! By the 1850s, America was well on its way to a confrontation with destiny as the South put all of its agricultural eggs in one basket and focused on the single crop of cotton. By 1860 all the ingredients for jump-starting the second great push in farming came together from all points in the Western Hemisphere.

WE'RE BACK!

The end of World War II came to a sudden halt in 1945 with the dropping of the atomic bomb. Actually, it took two of these horrific events to convince Emperor Hirohito and his advisers that Americans were dead serious about not invading Japan. Following the surrender, American troops mustered out of the military by the millions and integrated as best they could back into civilian life. During the war, food production had predictably gone down as the sixteen- to twenty-nine-year-old labor force went "over there." In the movie *Saving Private Ryan,* based on a true story about four brothers who fought during World War II, the issue of farm labor shortages that war creates was a featured theme. Three of four Niland brothers were killed in combat. All of them had come from a farming family in Minnesota. The government, upon learning of this tragedy, issued an order to save the last one by removing him from active duty and shipping him back home. They reasoned that if he were to die, yet another farm might go under from lack of manpower. During the war, the country tightened its belt and endured life with a food menu that had a greatly reduced number of choices. Americans ate less dairy and beef, and consumed far more starch and . . . well, more starch.

To compensate for the diversion of fresh and value-added produce to the military, ordinary citizens began growing their own food. Victory gardens sprang up everywhere conditions permitted, and the result was that Americans relearned the value of a freshly picked ripe tomato. The war created another unintended agricultural opportunity, too. American supply convoys destined for the Pacific Theater of Operations encountered Japanese submarines, and the U.S. ships were sitting ducks. Huge quantities of supplies of all kinds were lost, including K rations. It's not obvious who got the idea first, but many a commanding officer issued the order to establish hydroponic facilities on several of the captured islands. It is estimated that during the war, as much as eight thousand tons of fresh veggies were made available for Allied troops this way, helping to offset the losses suffered at sea.

After the war, hydroponics was discarded, forgotten entirely, as America turned its attention to growing crops on the land that was to become the site of the greatest agricultural initiative on earth. The baby boomers were now in their early teens and more interested in going to college than in staying down on the family farm. Apparently, this had little effect on the restarting of the U.S. food "machine." Despite all its attendant horrors, the war had proved advantageous in many respects, including supplying the right moment for the innovation needed to fully mechanize an entire country's armed forces. One group in particular was outstanding in this respect: the Construction Battalions known to everyone as the "Seabees." This naval group was highly decorated for its clever and efficient use of construction technologies. Using cutting-edge earth-moving machinery perfected during the early phases of the war, the Seabees rapidly adapted them to the efficient conversion of native jungle into barracks, and even small cities. They became particularly adept at remodeling even the most hostile terrain into flat airfields. After the war ended, these same methods were applied to land that before the war was considered highly unsuited for any kind of agriculture, easily transforming it into productive farmland using mammoth earth movers and massive tractors fitted with all kinds of front-loading devices. Farming was coming of age in the new industrial-military complex of postwar Europe and the United States. Good-bye wetlands, hello corn and cotton fields. In filling in many of the swamps in the American Southeast, we also got rid of malaria for good in this country.

NATURE ABHORS A VACUUM

At the same time that the war was raging in Europe and the Pacific, quietly, without notice, the Joad family's dust bowl was morphing back into its original ecological self: a series of tall- and short-grass prairies. It took only ten years for that wasteland to recover from the abuses of wheat and corn farming. How could this have happened? The riveting newsreel footage of 600-foot clouds of topsoil blowing across the plains states had convinced everyone that this was no place for humans to live, perhaps ever again. Yet, shortly after the mass exodus West, the wildlife returned. Seeds of native plants germinated and reestablished the tight-knitted, foot-deep root systems needed to hold in water from the odd rainstorm, and of course helped to preserve and restore the soil itself. Once the prairie grasses became the dominant plant species, all of the prairie's animals came out of hiding, too. Kit foxes, prairie chickens, burrowing owls, prairie dogs by the millions, even antelope and small isolated herds of wild buffalo and longhorn cattle could be found wandering the restored landscape by the 1950s. Evidence of nature's resiliency was everywhere. The depressing, lifeless landscape of the 1930s had returned to a good portion of its former glory, and without much in the way of assistance from us humans. In fact, it was because we left it alone by going off to fight the good fight that it was able to reach into its deep reserves of seeds buried under the sun-parched dust and rejuvenate a water-starved part of America.

BIG FOOT

Today, the world is getting hungrier by the minute. Basic nutrition, 1,500 calories of disease-free food, is already considered a luxury in some parts of Africa and India. Riots and hoarding routinely follow news about crop failures and projected shortages, especially when rice is the crop. The black market abounds with contraband rice and other essential grains, too, often confiscated from nongovernmental organizations (NGO) relief efforts to attempt to alleviate death by starvation. Darfur is a prime example of this current chaotic state of affairs. As alarming as all this sounds, it's conceded by almost everyone, including the most pessimistic of the agro-critics, that the world is still in pretty good shape in terms of the amount of food produced. It is so ironic that just to state this fact sounds like someone made the whole thing up, but alas, things are what they are. According to the Food and Agriculture Organization of the United Nations (FAO), food has never been more available than it is now. The USDA agrees with this.

We live in a world filled with inequalities and injustices so egregious that we can hardly bring ourselves to think about them. All the while, the earth's rapidly changing climate continues to point the way to an unprecedented upheaval in just about everything, but especially about where we can and cannot grow our food. Over the next twenty to thirty years we humans will experience a transition period in which established, proven agricultural practices will no longer be able to meet the needs of a rising population. Just look at what we have already carved out of virtually every terrestrial ecosystem for our farming needs: a landmass equivalent to the entire continent of South America, an outrageous agricultural footprint, and not just from a land-use perspective, either. Almost all farming requires some form of irrigation, and on a global scale uses around 70 percent of the available freshwater to do so. What suffers most in this case is the availability of drinking water. Water that's free of infectious diseases and toxins is becoming scarce in many places, especially where drinking water was already at a premium. In some water-challenged countries a barrel of drinking water is now valued higher than a 55 gallon drum of crude oil. In the process of irrigating fields, farming spoils the water it uses by producing runoff laden with salts and a whole catalog worth of agrochemicals applied in vast excess of what the plants really needed. Runoff can also include animal and human wastes. Taken together, farm runoff in all its forms is by far the world's most damaging source of pollution, creating dead zones wherever major rivers empty into the ocean. While runoff is always a problem, it is an even greater problem during times of flooding. Predictions from climatologists warn that over the next forty years, flooding will likely become more frequent and more severe, and will occur in many places that have never experienced this kind of environmental disturbance before. Agricultural runoff has already trashed numerous estuaries, as it makes its way from rivers out to the open ocean. In fact, no other species has ever disturbed the earth as much as we have, not even the dinosaurs. With another 3 billion of us on the way, most of whom will live in less developed countries

(LDC), it is estimated that we will need to set aside another Brazil's worth of land (10^9 hectares) to allow food production to continue as it's practiced today if we are to meet their caloric needs. That amount of additional arable land simply does not exist. If people are starving, and indeed they are, then it's due entirely to issues related to maldistribution and mean-spirited politics, not actual crop shortages. That is largely because we have learned how to force every last radish, ear of corn, head of lettuce, strawberry, and everything else we grow from land that was never biologically able to do so without our help.

RIBBIT

In addition to killing off most of the world's nurseries for shellfish and crustaceans, the overuse and misuse of agrochemicals has led to widespread ecological disruption in other zones, too. Two related examples will illustrate the point, although many others would equally qualify. In numerous wetlands throughout northern Minnesota, an unusual phenomenon has been noted and tracked. For at least the last ten years, a growing percentage of frog populations, mostly *Rana pipiens,* have been overproducing hermaphrodites, ordinarily a rare occurrence in nature. The cause of this teratogenic epidemic, elegantly documented by dedicated frog researchers including Dr. Tyrone Hayes of the University of California, Berkeley, could have been certain chemicals known for their ability to interrupt frog development. In the end, only one emerged as the guilty toxin: atrazine, a commonly used herbicide. Atrazine is widely used to control a variety of weeds that compete with crops such as wheat and corn. Runoff into wetlands, particularly in the upper Midwest of the United States, has had a major negative effect on frog and largemouth bass development, causing huge numbers to convert to hermaphrodites (i.e., animals that have both male and female sex organs). Other effects of atrazine include inhibition of the immune system, placing frogs and other amphibians such as salamanders at higher risk from infectious diseases caused by trematode parasites. In April 2006, the European Union banned the use of atrazine. As of this writing, the United States has yet to follow their example.

The atrazine problem has sounded the alarm as an example of the unintended consequences of using an agrochemical whose mode of action was either not fully explored or just ignored by the manufacturer. In a series of related studies carried out in California, leopard frogs with deformed limbs or extra limbs were found to be present in unusually high numbers in ponds situated within intensive agricultural settings. They were examined for potential causes and given a complete toxicology screen. A particular species of trematode parasite (*Ribeiroia*) was found to be the only common feature among many other possible causes. Researchers further determined that frog susceptibility to the infection was increased by prior exposure to atrazine, which adversely affected the frog's immune system. This herbicide is widely used throughout California in many agricultural situations. In the laboratory, the parasite was shown to cause leg deformities in a dose-dependent fashion. Interestingly, the parasite affecting California frogs has yet to be isolated from similarly affected frogs in the Midwest. More work is obviously needed.

The above findings are reminiscent of Rachel Carson's original call to arms in the 1960s, in which she identified DDT as the culprit. In her chiding of the agrochemical industry in her book *Silent Spring,* Carson did not condemn them entirely for producing vast quantities of DDT; rather, she mainly took to task those who used that powerful chemical in ways for which it had never been intended. Today, we find ourselves coming full circle to yet another catastrophic head-on collision with the same industry. When will we learn to use our intellectual gifts wisely and in a focused manner to spare those life forms in nature that we value by fully testing each and every agrochemical before unleashing them on the farm? Apparently, the earth can be used as a guinea pig for new agrochemicals until harm is proven, while humans are protected from misuse of drugs by stringent standards that govern the pharmaceutical industry, which is so regulated, that it now takes on average ten years before a new chemotherapeutic agent becomes commercialized.

Enough negativity for the moment. We should probably pause here and reflect on what we have accomplished, regardless of the ecological price we have had to pay for inventing methods to achieve a predictable food supply. These methods have allowed the human population to rise to 6.7 billion strong as of June 2009. With enough of us here now, we should be able to collectively rethink how we can get out of this mess and get on with our lives more in harmony with the land.

THE HAVES

Every bit of today's farming in the developed world is driven by technology. New kinds of farm equipment, new planting strategies, global-positioning systems for microcrop selection, and other high-tech approaches have extended the life of most soils far beyond their natural carrying capacity for producing, regardless of the crop in question. Research results from numerous graduate schools of agriculture throughout the world have led the way in changing what once was a hit-or-miss situation with regard to what and where we can farm into a predictable science of crop production. The agrochemical industries have been quick to adopt these new findings and have commercialized them to not only help the farmer but also to bolster their own profit margins. The Ag business is booming; fertilizer, herbicide, and pesticide product lines have become their bread and butter. No one denies the fact that these products have helped greatly in ushering in the second green revolution. Yield per acre for nearly every crop improves each year and will likely continue to do so for a few more years in many places. That is not to say that there are no problems looming on the near horizon, even for those fortunate countries with minimal environmental impact on their agricultural industries. But fossil fuels also figure into the equation. Higher yields are linked to increased fossil-fuel use by new farming machines. In the United States, more than 20 percent of all burned fuels goes to agriculture. The price of food is also linked to fossil-fuel use, and in 2008, the cost of food worldwide nearly doubled compared to the year 2000.

THE HAVE-NOTS

In stark contrast, things are quite different if one lives in a less developed country. As of 2009 there were forty-nine LDCs, most of which were in tropical regions of Africa and Central America. LDCs cannot afford commercially available fertilizers and are forced to use human and animal feces. This turns out to be the best way of spreading intestinal parasites from person to person. Worldwide, there are some 3 billion human infections with geohelminths (worm parasites of the gut tract transmitted through fecal contamination of the soil) that severely limit the health of an entire generation of children. Those LDCs that are located in tropical zones have poor soils to work with, save for those few countries in East Africa that are blessed with volcanically generated soils. It is well established that the vast majority of tropical soils are shallow at best, and cannot store significant amounts of carbon belowground. In addition, since it rains for a good portion of the year, valuable nutrients in the form of fallen leaves have to be recycled in days, as opposed to temperate-zone forests that may take an entire year to recycle their leaf litter into reusable nutrients. Last, tropical soils are poor in essential stored micronutrients because of leaching caused by the abundant rainfall. Growing significant amounts of food in these situations is impossible without nutrient supplementation, if not from feces then from the ashes of burned trees and shrubs that were cleared to make room for the crops ("slash and burn" agriculture). In this scenario, only three years' worth of crops can be harvested before the itinerate farmer has to move his family to yet another pristine forest site, where he then repeats the process. It is the single most common cause of deforestation in the tropics, with gold mining coming in a distant second. It is also the reason why in the tropics malnutrition is commonplace and starvation is routine, especially when a crop fails.

Crop failure can result from any number of things that are not factors in temperate climates. One that is foremost on the minds of all sub-Saharan African farmers is invasion by locusts. This voracious insect pest can actually smell the ripening of crops hundreds of miles away and seek them out before they can be harvested. The bottom line is that either the local affected population goes hungry and perhaps even starves to death, or they end up eating a diet heavy on locusts. Such was the case in Niger in 2003, when farmers there lost an entire year's worth of crops to the pest.

In the end, if and when our time on this planet has run its course, the human species will be judged not by the number of billionaires it has produced, or even by the exquisite art it has created over the entire span of its evolutionary history. Rather, it will be evaluated on how well it looked after its own kind and the rest of the life forms on which it was wholly dependent. Was the human culture based on equal sharing of resources in which every individual got enough water and food to live an adequately healthy life, or was it a species that encouraged greed and hoarding of resources for one group or country at the expense of others? We must answer this question in advance if we are to change from a destructive force of nature into one that has learned the true value of symbiosis. Learning how to provide for ourselves in every way—including producing food—in ways that do not encroach on the rights of others, such as the hardwood forest, will test our ability to think through the problem until we have solved it. I believe we can.

CHAPTER 4

TOMORROW'S AGRICULTURE

There is nothing wrong with change,

as long as it is in the right direction.

—WINSTON CHURCHILL

Rapid climate change (RCC) is the most important environmental issue that we face today, and it will continue to command our attention for the foreseeable future. It is literally rearranging the terrestrial landscape, and in ways that we have only recently begun to decipher, thanks to a new generation of satellites designed to measure things such as sea surface temperature, cloud formation, and droughts that affect crop production (see the images at http://earth observatory.nasa.gov).

Climate change affects every living thing on earth. Of particular concern is how RCC will affect our ability to farm. Where we grow crops today will be quite different from where we can grow them tomorrow, because patterns of temperature and precipitation are in a heightened state of flux. Greenhouse gases have added their own influence to these changing patterns and unfortunately will also contribute to the warming of the earth's atmosphere over the next hundred years, even if we were to stop using fossil fuels today, which of course we cannot. The carbon dioxide and nitrous oxide components of burning fossil fuels have also had a significant effect on the world's oceans, resulting in a more acidic environment. If this trend continues, the ocean's crustaceans, mollusks, and coral reefs will be in big trouble, since calcium carbonate, a major component of their shells and matrix, cannot form at pH values much below 8.0. The pH value of the ocean is now at 8.06; just twenty-five years ago it was 8.16.

The "direction" of climate changes favors those life forms that can most quickly adjust to them. Researchers used to think that plants would be the most severely challenged when it came to moving to a new comfort zone, due to the simple fact that they have no visible means of locomotion. Recent evidence seems to contradict this belief. For example, trees can relocate over great distances by seed dispersal mechanisms (wind, floods, birds, insects), and can germinate in a variety of places, some of which, even in a rapidly changing environment, will match exactly with their optimal ecological tolerances. Farmers are unfortunately wedded to the land they till, so when growing conditions change, favoring one country over another, the growers most adversely affected are those who have to remain in their own environment. It is inevitable that conflicts will flare up when more optimal growing conditions shift over the border into an adjacent country. Conflict has already occurred over the scarcity of water in many places. For example, in 2008, during the height of the twenty-five-year drought plaguing the Southeastern region of the United States, the states of Florida and Alabama sued the state of Georgia for withholding water releases from Lake Lanier. Atlanta gets its drinking water from that lake, but the Chattahoochee River below the hydroelectric dam eventually courses through parts of Florida and Alabama on its way to the ocean. The argument over the amount of released water considered minimally essential to satisfy the needs of those living downstream is, as of this writing, still in court and unresolved. Ironically, the following year, Georgia experienced massive flooding in September, in which an enormous quantity of topsoil was lost. Perhaps Georgia should charge Alabama and Florida for the additional water and soil it gave up to them.

The definition of how much is enough is largely determined by the haves and not by the have-nots. It's been widely speculated that the next war in the Middle East will be over water, not religion or oil. Food shortages now exist in many places, but as explained earlier, those are mostly due to the poor distribution of food, not the lack of it. Of course there are exceptions, and India was one of those when in 2006 it was forced to seek assistance from Canada and Australia in procuring enough wheat to meet the minimal needs of its people, who make up one-fourth of the earth's population. India's wheat harvest was severely reduced that year when a freak set of thunderstorms in the Punjab produced hail that knocked the grains off the ripened spears of wheat, rendering much of the crop unharvestable. In another part of the world and in that same year, rice blast—a devastating fungal disease of the rice plant, whose spores are dispersed by wind—created temporary food shortages in many parts of Southeast Asia, inducing food riots and hoarding of rice. Even the United States felt the repercussions from that farming disaster, and for several weeks rice was rationed to one 50 pound bag per customer per week at most of the nationwide discount food store chains. Did climate change cause these events, or were they just random perturbations in the weather? One thing is certain: If these events become routine, then we will have no choice but to attribute them to RCC.

Adjusting to a changing climate is what nature is all about, through a genetic process in which a mutant is selected from among many others in a given species that by mere chance turns out to be the fittest one for that particular environmental change. It's nature's hedge against extinction caused by the sudden rise of a set of adverse conditions. One can imagine how this might manifest itself on a string of interconnected volcanic islands in which change can occur moment to moment. In fact, it was just such a natural setting that triggered much of Charles Darwin's thinking when he visited the Galápagos Islands. By tracking the distribution of finch species, each of which had a different strength of beak, he was able to match them up with their food sources, a wide variety of plants that produced nuts and seeds of varying hardness. The stronger a bird's beak, the more likely it was that the bird fed on nuts and seeds that only its beak could crack open. In this way, the finches were able to evolve into sets of related species, occupying the same island and reducing the competition for food.

The process that produces these kinds of closely related species is known as adaptive radiation. As a given mutant plant species by chance produced a harder seedpod, avoiding overharvesting by finches with beaks too weak to open them, a mutant finch with a slightly stronger beak could take advantage of this untouched food supply, giving it a distinct advantage. In this way all of the finch species arose from a common ancestor, the original colonist, when the Galápagos Islands were just a single landmass. The oldest island is around 5 million years old, while the youngest is only 2 million years old. Darwin had the advantage of actually seeing all the different-aged islands during his five-week stay, and concluded correctly that nature has the ability to adjust to a rapidly changing environment by the process he came to refer to as natural selection. What he identified was nature's version of the arms race, expressed in a somewhat more peaceful manner. Today, scientists employing state-of-the-art molecular biological methods have confirmed Darwin's original observations on the link between beak and seed morphology, proving that in each case, single genes are all that is necessary to account for the changes in both beak strength and seed hardness. In fact, scientists have even been able to identify the original founder finch species. Indeed, how remarkable and magnificent is nature.

Why is it important to understand the fundamental biological process of natural selection? It's because we have gone down a different evolutionary path, in which we have employed artificial selection, as opposed to natural selection, to generate all of our food-bearing plants. None of them resembles the wild plant it was derived from. Our food crops have been picked and fussed over by generations of farmers, who began this journey some ten thousand years ago. They tailored them for characteristics that favored ease of cultivation and maximum yield of edible parts.

Corn is a good example of what happens when we discover an accident of nature and then capitalize on it (see chapter 2 for more details). Corn (*Zea mays*) started out as a small, inconspicuous grass in south-central Mexico. By chance, a mutant arose that had larger-than-normal kernels. It was easy to grow and had obvious nutritional advantages, so it quickly became adopted by the native tribes in the area of the Balsas Valley around 8,700 years ago. They nurtured and coaxed it into a plant that became totally dependent on irrigation for its survival. Increased water meant that maize (old-style corn) could spend more energy making reproductive structures (kernels). Over time, perhaps several hundred years, maize became fully domesticated. It was a sustainable, storable food source that grew wherever water could be provided, and it rapidly transformed cultures up and down the North and South American landscapes. If today's hybrid corn, the kind typically grown throughout the American Midwest, were to become reintroduced into the same environment in which it arose some nine thousand years ago and then left to its own devices, it would surely die.

It is obvious, therefore, that the traits we value as consumers of plants have little or nothing to do with the ones that enabled their ancestors to withstand severe environmental changes associated with droughts, floods, plant diseases, insect pests, and wide fluctuations in temperature. In short, we have bred the "wildness" out of them in favor of things that favor their growth as irrigated, pampered, well-fed monocultures, or "cultivars." If an environmental change should occur that exceeds the narrow tolerance limits for a given cultivar, then that crop will surely fail in that environment. It would be like turning our pet lap dogs loose in the woods and expecting them to survive on their own. Today in the United States, more than 90 percent of all seeds used in large-scale agriculture, regardless of the crop, are produced by just three companies. They are all highly domesticated strains with very narrow tolerance limits for temperature and precipitation. More than half the world's farmland is suboptimal with respect to most commercial crops and will only get worse over time. This is especially true throughout the tropics, in which the soils are mere inches thick and do not have much in the way of stored nutrients. The tropics are also the region of our planet experiencing the highest rates of population growth.

Throughout the ages, most of our crops (with the possible exception of wheat) have been carefully selected for growth in a series of narrow climate regimes. For this reason, regions of the world have become known for producing certain crops, but not all of them. It's worth emphasizing again that the conditions that plants depend on most relate to annual patterns of temperature and precipitation. Corn, rice, potatoes, and a wide variety of garden vegetables such as tomatoes, lettuce, cabbage, wine grapes, and the like, all have their own sets of ideal growth conditions. That is why most varieties of domesticated rice come from wet, semitropical climates, while potatoes do best in colder, harsher environments.

When ideal conditions for a given crop species occur

over a single growing season, farmers rejoice, and yields approach the plant's theoretical limit of productivity. Harvests are plentiful and the gods of agriculture are praised by all. However, when something such as a plant disease, insect pest, hailstorm, severe flood, or prolonged drought takes over, then all can be lost, sometimes overnight in the case of a killing frost or windstorm. Often, the only recourse is to plow the crop under and begin again next year. It is unusual to experience ideal growing conditions for more than a few years in a row, no matter where on earth we choose to observe them. Just ask any winemaker. This is the basis for vintage years. Rare indeed is the production of two consecutive years of great wine from the same vineyard, regardless of the growing region or the variety of grape. Quite often, however, most vintners manage to sell most of their annual production, vintage or no, but at much reduced prices compared to their best years. Most would agree that farming is not the easiest way to get rich, or even to scratch out a living in places where the conditions are marginal.

Environmental agronomists predict, based in large part on recent data regarding RCC, that crop failures will become more frequent in places in which they are now considered rare, and will become the rule rather than the exception in places in which they now regularly occur. Unfortunately, in this case forewarned is not forearmed. We simply don't have the luxury of time to breed into our cash crops the characteristics they will need to survive in an increasingly harsh environment, even though we think we know how to proceed in the laboratory. Modifying plants to resist droughts, ward off new plant diseases, and avoid attack from insect pests takes time and lots of funding, not to mention social acceptance for genetically modified plants. The environmental changes are just too global and rapid to expect much progress in these disparate areas over the next twenty-five years.

To further put things into perspective, *The Stern Review on the Economics of Climate Change,* published in 2006, estimated that over the next thirty years, RCC will cost the governments of the world a combined total of around $74 trillion. Expenses associated with rising ocean levels, significant loss of crops, increases in vector-borne disease transmission like malaria and the West Nile virus, and increased health care costs associated with these catastrophic events will consume the lion's share of federal budgets. Little will be left for social innovation, let alone critical research on new drugs, vaccines, and other life-saving technologies.

With farms around the world experiencing increased loss of yield due to more widely fluctuating patterns of weather, it's no wonder that in most developed countries farm insurance is a regular addition to their national budgets. Providing food for their citizens is an essential countrywide activity that benefits everyone, so farming must be able to continue next year even if the crops failed miserably this year. Subsidies are essential to keep the process moving forward. For example, in 2008 the U.S. Congress approved the Food, Conservation, and Energy Act to the level of $288 billion. This enhanced version of the 2003 Farm Bill included a healthy slice of that pie earmarked for farm insurance, and subsidies favoring home-grown crops over similar crops produced in other countries. The majority of the remaining funds was destined to pay U.S. farmers to grow food for agriculturally challenged countries. There were few congressional dissenters on the 2008 bill, as every state stood to benefit. On the other hand, the World Bank, the United Nations, the World Trade Organization, and other international governing bodies have repeatedly criticized the United States and other developed countries for creating economic situations that prevent fair trade of agricultural products, particularly those originating from less developed countries. While protecting a country's own farmers from competition and guaranteeing them a predictable income, such a short-sighted economic strategy prevents those disadvantaged nations from economically advancing into the more developed world. It's a valid point that needs addressing if we are to eventually behave as a true global community.

Yet even some of the most advantaged countries with respect to agriculture have experienced severe droughts and floods over the last five years, which has raised awareness as to the need to develop better strategies for food security and safety. More frequent outbreaks of food-borne infectious agents from organisms such as salmonella, cyclospora, and *E. coli* 0157:H7 have created a sense of urgency for the need to regroup and incorporate alternate technologies for growing food that is both safe to eat and locally produced, and hence more controllable from a food safety perspective.

The 2009 recall of all peanut-based products in the United States was the largest and most expensive recall of any food product in its history, including recalls involving ground meats contaminated with *E. coli* 0157:H7. The future of agriculture, at least as practiced outdoors, looks grim. Three examples will serve to illustrate just how much trouble we have created for our children and their children over the next twenty-five to fifty years if nothing changes.

John Steinbeck, had he lived long enough, would be truly amazed to learn that the migration West from the dust bowl created yet another more intractable agricultural dilemma, one totally unrelated to the labor movement or the civil rights of migrant farm workers. In the 1950s an agricultural strategy was adopted by a wide swath of farm cooperatives that would eventually irreparably trash an entire region of the Golden State, as large-scale farming took hold in the Central Valley of California and never left. A medium-size city's worth of migrant workers from Mexico joined the scene, and crops began appearing on the supermarket shelves of the world. The Central Valley became known as the new Garden of Eden for truck crops of all kinds: nuts, citrus fruits, vegetables, and table grapes. Industries that specialized in value-added products such as catsup, tomato sauces, and canned fruits all became ensconced in California. Hunt, Del Monte, Heinz, Dole, Sun-Maid, and other giant food producers made their fortunes there. Together with the dairy industry, agriculture has mushroomed into a $65-billion-a-year industry, with Salinas as its unofficial capital. All this growing requires huge amounts of water. The annual spring snowmelt from the Sierra Madres in California was not enough, and California growers petitioned the state of California to purchase water rights from the Colorado River, so they did. Snowpack in the Rocky Mountains averaged some 20 feet annually back in the 1980s and the state of Colorado had no particular use for the excess after it took its share, so it willingly signed away its water rights for twenty-year periods of time (for a price, of course). Massive irrigation schemes were quickly constructed to take advantage of this situation.

Soon, California water was abundant and cheap. Flood irrigation was the rule for many crops, including the almond industry. It was easy to implement and economical, too.

The Central Valley is one of the hottest, driest places in America, with average daily temperatures in the summer approaching 110°F. Ecologically, the majority of the land was a mixed-grass prairie completely surrounded by high mountains; hence the dry, hot environment. It's about the least likely place to attempt to farm in North America without heroic assistance from hydroelectric projects. In addition to needing extra water, since the soil types are nutrient-stingy and favor semidesert plants, almost all of California's main crops require lots of fertilizers. As crops became the norm, uninvited guests—insects and weeds—prompted the heavy application of a new generation of pesticides and herbicides. Year after year, through the 1960s, '70s, and '80s, California growers boasted the highest per capita income compared to any other farm group in the world, attracting even more agriculture to the Central Valley.

Then one day, a funny thing happened. Actually, no one laughed. Ponds started to form by groundwater welling up from saturated aquifers, creating an artificial wetlandlike ecology in many localities throughout the 300-mile-long valley. Migrating waterfowl began using these newfound bodies of water to nest along their shores. But things were not what they seemed to be. Birds of all species soon began to die in large numbers. Testing revealed that many ponds were contaminated with high levels of selenium and other heavy metals, trace components of fertilizers that were being used indiscriminately by nearly every large farm operation. Pesticide levels were high, too. Most disturbing was the water's high salt content. The toxic groundwater had reached the bedrock below and was now headed up toward the surface; each year's flood irrigation activity brought it that much closer to the taproots of almond and citrus trees.

One study published in the *Proceedings of the National Academy of Sciences* in 2005 gave a detailed outlook for the next twenty-five to fifty years based on current practices. The outlook was positively depressing. According to this study, failure of much of the southern half of the Great Central Valley is certain due to high salt levels in groundwater, unless another method of supplying water to crops can be implemented that does not allow runoff to accumulate in the aquifer. But finding such a method may not be possible if California runs out of water, which it might over the next twenty to fifty years. The Department of Energy secretary, Nobel Prize winner Steven Chu, flatly stated three weeks after he took office in 2009 that the entire agricultural sector of California would become obsolete in less than fifty years due to lack of a source of noncontaminated fresh water: "I don't think the American public has gripped in its gut what could happen. We're looking at a scenario where there's no more agriculture in

California." PBS's Bill Moyers produced a *Frontline* story in 1993 titled "In Our Children's Food," documenting the environmental disasters that were incubating just under the surface of the most productive agricultural site on the planet. If the Central Valley fails, then consumers the world over will feel the economic effects, as food prices will certainly go even higher than predicted based on rising fuel prices.

The island of Chongming off the coast of Shanghai "grows" at a rate of about 150 meters per year. This may not seem like much, but consider that it is roughly 40 miles across and the second-largest island in China, and one gets the feeling that something unusual is happening. Indeed, there is. Valuable topsoil, farmland quality, is being transported by the mighty Yangtze River from that country's heartland to the edge of the South China Sea due to an annual flooding regime driven by RCC. Deforestation to make room for more and more cropland has made the central region of China vulnerable to even moderate alluvial erosion, but in the last ten years, floods of millennial proportion have altered the way China farms. Even the Three Gorges Dam project cannot bring back the soil lost over that time period. In the seventh century, Chongming was only several sandbars wide and not much to look at. Then, as the population rose and agriculture took off, the island began to form, and within several hundred years Chongming was visible from outer space, along with China's world-famous 1,500-mile-long Great Wall. If you recently bought property at the tip of Chongming hoping to build a house with an ocean view, think again: Within just three years, your South China Sea villa would soon become surrounded by a dense forest growing on new land that was deposited right before your very eyes. Perhaps we will have to modify Will Rogers's prophetic statement about investing in real estate. The island is now so big that a planned city of some five hundred thousand people was contemplated, to be named Dongtan. Its purpose was to demonstrate how to live sustainably in a carbon-neutral way. Ironically, the project had to be scrapped due to the "nonsustainability" of funding from the Shanghai International Investment Corporation.

With the fertile countryside disappearing downstream and into the sunset, maybe the future of farming in China will manifest itself better and more efficiently within the urban environment. I had the privilege of attending a seminal meeting held in Beijing in 2007 at which the countries of the Netherlands and China gathered to discuss bringing farming to the cities. The meeting was titled "Innovating Metropolitan Agriculture." If perseverance and need are the driving forces for innovation, then China will win the day, hands down, as they have done for so many other challenges throughout their rich and long history.

The third and final example is somewhat less upbeat and

has ominous overtones of impending doom. Established in 1971, Bangladesh is a country that averages just 5 feet above sea level. It is home to some 154 million people and is only 133,910 square kilometers, or roughly half the size of the state of Texas, making it one of the most densely populated regions on Earth. Bangladesh is the recipient of all the waters flowing down the Ganges and Brahmaputra rivers, and both join up to then empty into the Indian Ocean through a massive, tangled network of estuaries and deltas. Flooding is a regular part of the country's ecological heritage. In 2008 the monsoons came on particularly strong and washed away topsoil that would have provided food for some 2 million Bangladeshi. That year, 2 million other Bangladeshi shared their meager larders with these unfortunate individuals. The flooding of 2008 was a single event that forever rearranged the farming landscape, placing even greater stress on that country's agricultural system. Bangladesh's next-door neighbor, India, is not immune to these events, either. The subcontinent experienced a massive flood in October 2009 in the state of Karnataka that wiped out a wide swath of farmland, killing thousands of farm animals and hundreds of innocent people in the process.

As RCC tightens its grip on the land, it's obvious that tomorrow's agriculture will become increasingly tenuous, and farming as an occupation will continue to attract fewer people to it. For example, in the United States in the 1930s, there were about 6 million people listing themselves as farmers (this figure included their immediate families, as well). In 2009 there were fewer than 150,000 individuals who declared farming as their principal occupation. At the same time, urbanization will continue to exert its pull on those living in the suburbs and rural areas. The single most important reason for people moving to the city, worldwide, has its roots in failed farming. Today, the designation "farmer" is not even listed as a separate occupation on the U.S. national census form, since there are currently so few of them.

As farmer numbers drop and as the size of the average farm increases due to industrialization of farm operations, conglomerates have developed their own set of problems, one of which is the cost of food inspection. For example, attempts to cut expenses related to sanitary practices have led to widespread outbreaks of a variety of foodborne illnesses. The public is no longer convinced that big agriculture has their best interests at heart, and this has resulted in a deep mistrust of mass-produced food. If the current global economic downturn continues to deepen, cutting costs wherever possible will become even more tempting to companies that are already at the edge of bankruptcy.

The topic of food safety has become so acute that it has risen to the top of the list of things that consumers want most in their food supply. As an outgrowth of this crisis, many advocate groups in urban centers throughout the United States and elsewhere have voiced their demand for produce that is free of agrochemicals such as pesticides and herbicides. The phrase "organically grown" has become the mantra of a new "foodie" generation of middle-class consumers, championed by the likes of Michael Pollan and Alice Waters. Numerous food retailers, Whole Foods in particular, have risen to the challenge, and have gone to great lengths to ensure a steady supply of organic produce on their shelves in an increasing effort to meet the burgeoning demand for what many perceive as healthier, albeit more expensive, food items. Portland, Oregon, has its own local food mart chain, New Seasons Market. Patterned somewhat after the Whole Foods model, New Seasons Market went public in 2000 as a single store offering choices that included organic, locally grown produce. By the summer of 2009, it opened its ninth store in Portland, this time in the middle of an urban food desert (if one can imagine such a thing existing in that food-conscious city), and captured the inner-city market with reasonably priced fresh produce that practically flew off the shelves the moment it opened. Within the last ten years, a groundswell of consumer interest in supporting local farmers has emerged, spearheaded by the Slow Food movement. The urban landscape has become the new agricultural frontier for rooftop vegetable gardens, restaurants growing their own herbs and spices, and city dwellers forming consortia for the production of all kinds of produce. Urban farming has caught on in places such as Portland, San Francisco, New York, Chicago, and Vancouver, Canada. If this trend continues to gain momentum, then, as with China, the next agricultural revolution might well be an urban-based one. But regardless of the reasons, it's apparent that something radically different is needed to allow for the large-scale production of locally grown food, especially within the urban landscape, as RCC progresses to the next level of an environmental crisis that appears to be unavoidable.

If farming as we know it collapses under the heavy weight of climate change, what will be the fate of the agrochemical industry? The production of ammonium nitrate–based fertilizers, herbicides, and pesticides has been the mainstay of companies such as Dow Chemical, Cargill, Archer Daniels Midland, Monsanto, the Mosaic Company, PotashCorp, and Agrium, Inc. The pesticide business is the most widely diversified of the three agrochemical industries. The National Pesticide Information Center, co-sponsored by the Environmental Protection Agency (EPA) and Oregon State University, lists no fewer than 443 companies that manufacture at least one major pesticide, as of the summer of 2009. It's no wonder that regulating pesticide use has become so difficult. Environmentalists have repeatedly lobbied for tighter controls over their use, but with only modest success. Rachel Carson would still be outraged over the lack of good practices for most pesticides currently on the market.

The laboratory manipulation of commercial crop plants, another new venture from the agrobusiness sector, has become highly lucrative for some plants, corn, for instance. Generally referred to as genetically modified organisms (GMOs), this initiative has come under attack because of a perception on the part of the public that GMOs are potentially harmful and should not be allowed. In fact, they have been modified to resist droughts, attack from a variety of plant pathogens, and increased amounts of herbicides. Nonetheless, the industrialization of crops has galvanized a number of advocate groups worldwide that oppose all genetic research on commercial produce. It's interesting to note that these same groups did not object strenuously to the introduction of a fast-growing, flood-tolerant "super" rice that was modified in the laboratory by Monkombu Sambasivan Swaminathan at the International Rice Research Institute in the Philippines. Today, over half the world enjoys a better life because of his pioneering work. Despite this beneficial research, not even GMOs can compete with nature and RCC. Similarly, over the next twenty to forty years, if even the most conservative predictions regarding the linkages between climate change and crop failures are considered, applying more and more agrochemicals to land that continues to give lower and lower yields will not solve the problem of how to feed the next 3 billion people. An ecological solution is required that takes into account all the life forms on our planet.

Can a city produce most of its own food and recycle most or all of its own wastes? I believe the answer is yes. In fact, I know the answer is yes. There are many new modalities for growing produce of all kinds indoors that will make any urban food production scheme possible. It's called controlled indoor environment agriculture. High-tech greenhouse farming is already being deployed in many places around the world, most notably in New Zealand, the Netherlands, Germany, England, Australia, Canada, and the United States. Hydroponics, aeroponics, and drip-irrigation methods have improved vastly over the last ten years, to the point of revolutionizing the ways in which we can produce indoor crops at will. Large-scale commercial indoor facilities located in Arizona, England, and the Netherlands have proven to be highly profitable. The only missing element is urbanization of the concept. Reengineering greenhouses from a horizontal footprint to one that conserves space by stacking them on top of each other is all that is needed to bring them into the city proper. Abandoned urban spaces can then be fully utilized. In addition, vertical farms of varying heights can be constructed to meet the needs of restaurants, school cafeterias, hospitals, and apartment complexes. Some stand-alone vertical farms will also surely be built for mass production of essential crops such as rice, wheat, corn, and other grains, even crops for the production of biodiesel. These might be located just on the outskirts of the city, where land is perhaps more available and/or more economical. Taken together, urban farming in specially constructed buildings specifically designed to grow whatever crops we need without the application of soil-based technologies is the founding principal upon which tomorrow's agriculture will be based.

Municipal waste management is another area that has improved by quantum leaps over the last ten years. Today, modern incineration strategies, including plasma arc gasification, are widely accepted for handling most kinds of wastes throughout Europe, and Germany has become a world leader in applying incineration to the safe disposal of both solid and liquid waste streams. Near pollution-free burning of solid municipal waste has all but eliminated the need for landfills, and at the same time produced much-needed energy through steam-driven power generation of electricity. Grey water and sludge derived from black water can also be handled in such a way as to allow any municipality to reap the benefits of recycling its wastewater, while extracting energy from the solids. The combination of all of these currently applied strategies will allow the city of the future to mimic the best qualities of an intact, functional ecosystem: bioproductivity and zero waste. Tomorrow's urban environment will be the focal point for the rest of this book, in which the details of such a scheme will be presented.

CHAPTER 5

THE VERTICAL FARM: ADVANTAGES

We must become the change we want to see.

—MAHATMA GANDHI

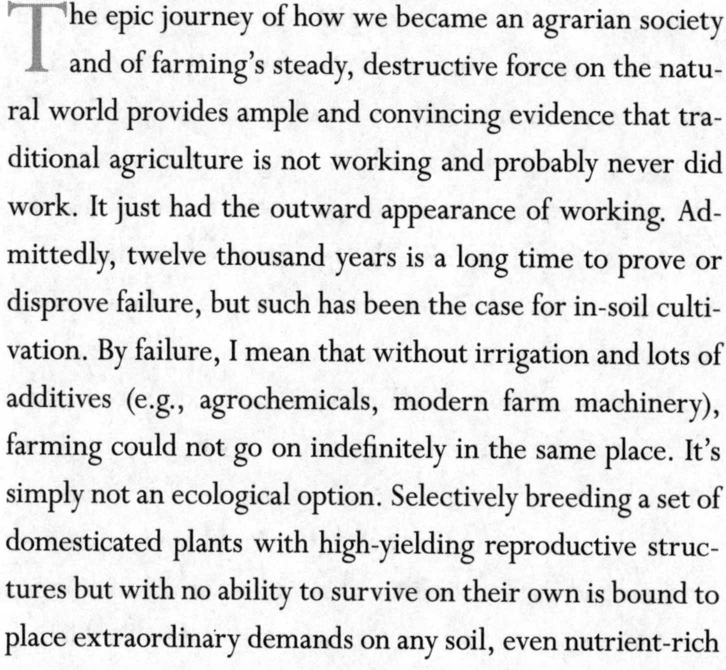

The epic journey of how we became an agrarian society and of farming's steady, destructive force on the natural world provides ample and convincing evidence that traditional agriculture is not working and probably never did work. It just had the outward appearance of working. Admittedly, twelve thousand years is a long time to prove or disprove failure, but such has been the case for in-soil cultivation. By failure, I mean that without irrigation and lots of additives (e.g., agrochemicals, modern farm machinery), farming could not go on indefinitely in the same place. It's simply not an ecological option. Selectively breeding a set of domesticated plants with high-yielding reproductive structures but with no ability to survive on their own is bound to place extraordinary demands on any soil, even nutrient-rich volcanic deposits. By contrast, monocultures are quite rare in nature, with some stands of trees and wide swaths of tall grass prairies being the primary exceptions. Even in those examples, though, there is ample diversity of other plant species, vertebrates of many kinds, and a myriad of invertebrates and microbes living on and in the soil.

Biodiversity means there is competition among all the assemblages of plants and animals for resources. It also means that there is much recycling of nutrients going on. For that reason, competition on farmland, if allowed to go unchecked, would greatly reduce yields. Sharing resources with the rest of the surrounding wildlife is not an option for commercial farming, no matter how "ecological" the farmer claims to be. Therefore, the sole purpose of agrochemicals is to reduce—or, better yet, completely eliminate—the competition, favoring the crop of choice by killing off the insects and unwanted weeds. It should also be noted that over the past fifty years of implementing this two-pronged chemical strategy, numerous weedlike plants have become more and more resistant to herbicides, while insect pests have become almost totally resistant to a wide variety of pesticides. It's natural selection at work, no matter how clever we get in the laboratory.

At the same time that our agricultural landscapes have been pushed to their limits and beyond, our population has just about risen to the point of no return—less land left to cultivate but still plenty of hungry mouths to feed, and 3 billion more on the way. We have no other choice but to conclude that farming on soil is not a long-term sustainable solution to meeting our population's energy needs, period. Environmental scientists predict that if things do not change, soils will soon collapse under the heavy burden of the application of too much short-sighted technology and not enough long-term ecological planning. Like the ostrich, we have buried our heads in the sand, hoping that this impending crisis may somehow go away. So we continue to deforest, to plow and plant more tracts of land, to irrigate using up more precious freshwater, and to spread more powerful—and more harmful—agrochemicals over the now biologically impoverished earth. When we use human feces as the fertilizer, as is the case in most of the less developed countries where commercial ones are too expensive, we also encourage the spread of numerous gastrointestinal parasitic worm infections. To get a feeling for how serious a global health problem this has become, a clinical correlation has recently been discovered in which four different helminth infections, often found together in the same individuals, can undermine the health status of those already infected with either HIV/AIDS, malaria, or tuberculosis: "Therefore, achieving success in the global fight against HIV/AIDS, tuberculosis, and malaria may well require a concurrent attack on the neglected tropical diseases (re: the hookworms,

Ascaris, Trichuris and the schistosomes) and waging a larger battle against a new 21st century 'gang of four.' " What's more, the elimination of this gang o' worms from the environment requires nothing more than good sanitation. We don't need any drugs, vaccines, or any other complicated, expensive medical intervention strategy; avoid contaminating food or water with human feces containing their infectious stages and they go away. That is why the Western world does not have much in the way of those diseases, though it still has to deal with sexually transmitted HIV/AIDS and droplet-transmitted tuberculosis. Thus, elimination of the use of human feces as fertilizer goes a long way to solving a major global health problem. Indoor farming would be an integral part of that solution.

NO MAN IS AN ISLAND

Like it or not, we cannot live out our lives apart from nature. Scientists from all of the subdisciplines of ecology have independently come to the same conclusion; namely, that all of life on earth is linked either directly or indirectly to each other in mutually dependent life-renewing cycles. It is the foundation upon which that science is built. Without our interference, life would go on in an equitable manner, with all the life forms living within a given eco-zone sharing their part of the energy budget provided to them each day by the sun. We have always been part and parcel of that scheme, but only recently in our history, just over the last fifty years or so, have we come to appreciate these intimate connections from a formal, scientific perspective. Today, we find ourselves looking around somewhat embarrassed, wondering why we have been so unfair to the rest of the life forms on the planet. Not all human societies have behaved that way, however. Many native cultures that depend solely upon nature for their livelihood have survived quite well without the benefit of modern technology, for example, the Australian Aborigines in Arnhem Land and the Yanomami of Brazil. They learned to live in balance with their environment by closely observing their surroundings, and have established long-term relationships with the life forms that comprise their world. To them, spoiling nature would be tantamount to committing suicide at the societal level. Destroying the very thing that keeps them alive would never even occur to them. Most natural cultures have no word for garbage or waste, for example. Many societies that did not realize this intrinsic relationship perished long ago, as author Jared Diamond describes in his book *Collapse*. Overharvesting and greed translate to extinction in any language. As some cultures grew in numbers, their biological needs increased disproportionately. They now had the capacity for advanced reasoning and creativity, and used these two intellectual attributes to invent farming, and eventually the rest of our technology-driven world.

CONNECTIVITY

Nonetheless, despite all of our cleverness, the connections between us and the rest of the world remain strong and immutable. The air we breathe, the water we drink, and the food we consume are examples of our ties to natural processes that help us survive, even today in the most sophisticated, advanced societies. Nature ensures that these things are made available and are safe to consume. In contrast, we who live in the techno-sphere have consciously chosen not to integrate our lives with nature, at the expense of the biosphere. Fortunately, we have not yet learned to directly control the hydrological cycle or any of the other biogeochemical cycles that ensure that matter gets reused. Once again, it's the science of ecology that has redefined these natural cycling processes in terms of services. They have both intrinsic and apparently economic value as well; it is estimated that all the ecological services on earth may be worth as much as $60 trillion. These services are sometimes referred to as "natural capital" by some who feel the need to put a dollar figure on the very processes that keep us alive. In my view, the things that life on Earth have evolved into—that reinforce processes that keep us all alive—are absolutely essential, and therefore to couch their worth to us in terms of money is crass and degrades the concept of life itself. It is obviously a character flaw in the human genome and, I would imagine, also highly offensive to the rest of the organisms who all share a common evolutionary ancestry with the first life forms on the planet. The need to reduce all of nature to an economic paradigm is indicative of our need to anthropomorphize virtually everything to conform to our worldview that we are at its center. Nothing could be further from the truth. How could we have so quickly forgotten the most important lesson of all, taught to us by none other than Copernicus: The universe does not revolve around us, and neither does the biosphere. Deny and/or ignore our relationships with the rest of the world and we will surely perish. That is the simple truth.

All of the available scientific evidence points to the fact that the most destructive force on earth is our penchant for encroaching into natural systems, mostly for the purpose of producing more food. But, locked into our present mode of food procurement, what choice do we have? We have indeed become trapped, held prisoner by our own device (shades of "Hotel California"), locked into an ancient, outdated system of food production that requires us to use more and more land to address the demands of a rising human population. If we continue down this dead-end road, then Malthus will indeed have been correct, if a tad premature in his predictions. Unless, of course, another set of technological breakthroughs once again comes to the rescue and pulls us off the tracks of impending doom. However, what is most required at this point in our history is not yet another quick techno-fix, but rather a permanent overhaul in the way we behave as a species. I believe there must be ways to satisfy both our needs and the needs of those creatures who do us no harm. I think that creating sustainable food-generating systems within the urban landscape would be an excellent first step, and would solve a number of problems associated with environmental destruction. A city-based agricultural system would allow us to carry out our lives without further damaging the environment. In fact, by relieving a sizable portion of the land of its food-production obligations, we would become two-time winners; we'd still get our food, and we would begin to regenerate the ecological services we unwittingly forfeited when we encroached into natural systems for the sake of our own benefit without any thought to much else.

Doomsday will have to wait, I'm afraid, barring some catastrophic hit from a Manhattan-size meteor or the sun unexpectedly going supernova on us, for what I am about to present in detail is a realistic, workable solution addressing food production and environmental repair. By applying state-of-the-art controlled-environment agricultural technologies as an integrated system contained within a multistory building—vertical farming—the world could rapidly become a much better place to welcome the next generation of humans. City life is what we are all about, and creating a balanced coexistence with the rural remainder is not only achievable, but highly desirable and economically viable. Considering the cost of addressing rapid climate change in the $70–80 trillion range (*The Stern Review on the Economics of Climate Change*), which is the total value of ecological services on this planet, and urban farming in tall buildings becomes, please pardon the cliché, a "no-brainer."

The advantages of growing the majority of our crops inside the city limits in vertical farms are listed in the table on pages 145–146. Many of them also apply to controlled-environment agriculture in single-story greenhouses. The differences relate to food harvesting, storage, and shipping issues, as well as to the size of their respective ecological footprints. Almost all high-tech greenhouses lie well outside the city limits, because, as pointed out, land is much cheaper there. The farther away food production gets from the urban center, however, the larger its ecological footprint. Undoubtedly there are many other good reasons for establishing vertical farms that will become apparent after a few are up and running. Over the years, I have brainstormed at length about the idea and have, so far, come up with no significant disadvantages, save for the initial costs of construction and the question of what to do about displaced farmers. Even here I think we have viable long-term solutions. For the farmers, I would lobby long and hard for a political solution allowing them to reap the benefits of sequestering carbon. Abandoned farmland rapidly returns to its ecological setting; witness the entire Northeastern portion of the United States. For an elegantly written treatise on this subject, I highly recommend *A Sand County Almanac* by Aldo Leopold, which documents in lucid, beautifully descriptive language how his father's farm in Sauk County, Wisconsin, grew back into a hardwood forest. Pricing carbon at its true value would create an economic incentive for farmers, most of whom are on the edge of eking out a living, to finally make a decent wage and at the same time help to restore damaged ecosystems. Allowing the trees to grow back would help slow down climate change by sequestering carbon from the atmosphere, and would also increase the biodiversity of impoverished, fragmented woodland. As far as the expenses incurred in the "invention" of a vertical farm, I would venture to guess that any first edition of an invention is going to cost a lot. As the invention becomes accepted and demand for it increases, the price of each one will go down. Take any one of our modern conveniences—air travel, the hybrid car, plasma screen televisions, the cell phone, the handheld calculator, for example—and you get the idea. I fully expect that vertical farms will succeed when we realize their true worth, not only for us, but for the rest of nature, as well.

In summary, implementation of the vertical farm employing large-scale hydroponics and aeroponics inside the cityscape is a potential solution for two problems: production of food crops to feed a growing urban population without further damaging the environment, and freeing up farmland and allowing it to return to its ecological setting. In most cases, this means restoration of hardwood forests.

ADVANTAGES OF THE VERTICAL FARM

1. Year-round crop production
2. No weather-related crop failures
3. No agricultural runoff
4. Allowance for ecosystem restoration
5. No use of pesticides, herbicides, or fertilizers
6. Use of 70–95 percent less water
7. Greatly reduced food miles

8. More control of food safety and security
9. New employment opportunities
10. Purification of grey water to drinking water
11. Animal feed from postharvest plant material

1. *Year-round Crop Production*

Since the beginning of agriculture, crop production has been linked to the seasons, even in tropical climates. The time of year and patterns of weather, together with the soil types found there, determine the yield of a particular crop in any region. Failure to produce maximum yields has traditionally been associated with adverse weather conditions that either arrive late in the growing season or are associated with reduced or increased amounts of rainfall. Such has been the case lately for the otherwise highly dependable monsoons. For example, for the last ten years, throughout most of India, the monsoons have been late in coming and too short in duration, but produce the same amount of rainfall. The difference between the past and the present, of course, is that now not enough water soaks into the ground to last the year, and floods are a regular occurrence. As a consequence, topsoil is being lost at an alarming rate, and crop failures abound due to water shortages near the end of the growing season. In many places throughout India, agricultural runoff is out of control. An intense increase in the rate of urbanization for all major urban centers in India, due in large measure to the migration of farmers and their families to the cities, is another unwanted consequence of an erratic monsoon season, and places an even greater burden on municipal services, many of which had already been stretched to the maximum and beyond. Other regions that rely on the monsoons for almost everything associated with water are suffering a similar fate.

The advantage of not having to be concerned with conditions outside is obvious to everyone. It means that a farmer can plan to grow any crop at any time, and anywhere. Not only is this a better, more reliable strategy for sustainable food production, but it also allows the farmer to take advantage of seasonal markets that may permit a crop to be sold at a much higher than normal price. A good example is the sale of tomatoes in Europe each year during the late summer months. When the crops are in season, trade agreements go into effect that favor the sale of local produce. When sales drop off as the season progresses, tariffs go down. This is when controlled-environment agriculture shines, since a farmer in Morocco, for instance, can plant hydroponic tomatoes to mature at just the right time to be able to sell them at their highest prices in Spain, extending the season for tomatoes there until the next year, when the tariffs once again go into effect.

2. *No Weather-related Crop Failures*

Indoor farmers do not have to pray for rain, or sunshine, or moderate temperatures, or anything else related to the production of food crops, for that matter. That is because they get to control everything: the temperature and humidity, as well as the amount of light and the density of the plants. Over the last few years, there have been catastrophic weather events on a global scale that have permanently altered the way food is produced. Floods, droughts, tornadoes, hailstorms, cyclones, hurricanes, and high winds are some of the reasons why outdoor farming is a precarious occupation at best. In the United States, hurricanes, protracted periods of rain, and droughts have been the main villains. On August 24, 1992, the category 5 Hurricane Andrew ravaged the lower one-third of Florida and left $34 billion in damage in its wake. Most of it was property related, but there was also a significant portion of farmland destroyed. Florida is the second-largest cattle producer in the United States and one of the largest sugarcane growing regions in the Western Hemisphere. The following list of produce was supplied by the State of Florida Department of Agriculture for the year 2005:

56 percent of the total U.S. value of production for oranges ($843 million)

52 percent of the total U.S. value of production for grapefruit ($208 million)

53 percent of the total U.S. value of production for tangerines ($68.4 million)

53 percent of the total U.S. value of production for sugarcane for sugar and seed ($433 million as of 2004)

49 percent of the total U.S. value of sales for fresh market tomatoes ($805 million)

44 percent of the total U.S. value of sales for bell peppers ($213 million)

31 percent of the total U.S. value of sales for cucumbers for fresh market ($73.7 million)

31 percent of the total U.S. value of sales for watermelons ($127 million)

While many well-off Florida farmers were able to recoup their losses via crop insurance, some who were most affected by the storm decided to scrap traditional planting and harvesting methods and reinvent themselves as indoor agriculturists. One strawberry farmer who wishes to remain anonymous decided to replace his now destroyed 30 acre farm by constructing a high-tech greenhouse with a 1 acre footprint. Using hydrostackers, he was able to produce the equivalent of 29 acres' worth of fruit, with year-round production. He elected to return the rest of his farm to its natural setting by simply leaving it alone. Within two years, the understory had returned and the biodiversity of the land improved dramatically. It's instructive to remember the old axiom: "Nature abhors a vacuum." Perhaps this farmer's main concern now will be how to keep the alligators and water snakes out of the family swimming pool. Everybody, including the wildlife, was finally happy!

As previously discussed, flooding has become a chronic problem throughout much of Southeast Asia and the Indian subcontinent. Droughts, too, have recently taken their toll on agricultural productivity, especially in sub-Saharan Africa, the American Southeast, and Australia. Without water to irrigate, farming always fails. Floods and droughts result in loss of topsoil, too, the world's second most serious agricultural problem after the toxic effects of runoff. Replacing lost soil by natural processes takes years. Indoor soil-less crop production using water-conserving hydroponics is the only reasonable approach to avoid this outcome.

3. *No Agricultural Runoff*

The USDA unequivocally states: "Agricultural nonpoint source pollution is the primary cause of pollution in the U.S." (http://www.ars.USda.gov/). Even if flooding is not considered, significant runoff from farming still occurs as the result of the majority of irrigation practices (the exception being drip irrigation). Runoff is essentially unpreventable, given the fact that in order to maximize yields with conventional outdoor crop production, almost every plant species requires more water than the amount they receive from rain events. Runoff in most advanced farming operations is laden with silt, fertilizer, pesticides, and herbicides, and usually ends up in some river on its way to the estuary. In both of these aquatic environments agrochemicals take their toll on wildlife, including mollusks, crustaceans (shrimp and crabs), and fish. The nitrogen portion of fertilizers scavenges oxygen from the water column, creating a "killing field" for fresh- and saltwater organisms. For this reason, the United States is forced to import nearly 80 percent of its seafood. A similar situation exists in many other places where agriculture is dependent on the heavy use of these environment-altering compounds. What's more, those organisms that do survive the onslaught of pollution are certain to contain versions of many agrochemicals in their flesh due to bioaccumulation up through the food chain. This has raised havoc with fish and amphibians that breed in freshwater.

California is divided into three agricultural regions: north, central, and south. In the north, land in the Sacramento River basin produces 20 percent of California's revenue from agriculture and has always had chronic problems associated with urban as well as farm runoff. The Sacramento River supplies some drinking water to San Francisco, and to all other communities in the northern Central Valley, so concern in this case is not just for what happens to the wildlife of the Sacramento estuary. The biggest offender appears to be a pesticide, diazinon, employed as a generic insect control agent on a variety of crops. Coalitions of concerned residents who live within the watershed of that river have brought political pressure to the state government to improve the monitoring of the entire drainage basin. It is an ongoing battle that has public health as its main concern.

The lower half of the agricultural region of California is divided between the central and southern zones. In the south, the Colorado River is diverted to supply water for irrigation in Arizona and California, and then what's left of it empties into the Gulf of California. The central zone is much larger in area and in crop production. Eleven of the top twenty California counties for agriculture are located there, and the San Joaquin Valley has eight of them. The

Central Valley derives its water from many rivers that originate in the Sierra Nevada Mountains. Several of them terminate in lakes within the Central Valley, or connect with the Sacramento River. However, none of the rivers in the southern portion of the Central Valley flows through its landscape to the Pacific Ocean, which creates an unusual situation with respect to agricultural runoff. As detailed earlier, the irrigation water simply soaks into the ground. It has nowhere else to go but down. Eventually, à la some Edgar Allan Poe horror story—"The Cask of Amontillado" is the one that comes to mind for me—when the toxin- and salt-saturated water table eventually rises to the level of the deepest taproots, the death knell for the entire region will have sounded, and agriculture as we know it will cease to exist in Steinbeck's land of milk and honey. If current irrigation practices continue for another twenty-five to thirty years, California will begin to feel the effects of this impending disaster. Losing all of its crop-growing potential could cost that state as much as $30–50 billion in annual agricultural revenue. All of the damage caused by runoff can be prevented by shifting to an indoor cultivation strategy. The water used to grow food inside could even be recirculated and used again and again, provided that nutrients are replaced at the same rate that they are taken up by the hydroponically grown plants.

As documented earlier, the growth of Chongming Island

off the coast of Shanghai from three insignificant sandbars to China's second-largest island in just several hundred years is further evidence of the power of agricultural runoff to permanently alter the landscape.

4. *Allowance for Ecosystem Restoration*

If a significant amount of farming were to take place inside the urban landscape, then the world's ecological footprint of agriculture would become smaller. For most crops, about ten to twenty times the acreage it takes to raise them indoors could be converted back into hardwood forest outside. This is because crops can be grown year-round and none would be lost due to severe weather events. Large-scale environmental restoration is high on everyone's list of things we wish we could do, but most perceive it as an unrealistic goal due to the amount of land we now need to farm, with more land needed in the near future as the human population continues to increase. The Food and Agriculture Organization laments in each edition of its *State of Food Insecurity in the World* reports that the simple solution to restoring the natural world is to leave it alone. One could seriously question this strategy of benign neglect, since the great majority of us now live either in cites or their suburbs and have never had the occasion to witness nature in a restorative mode. But be assured that there are numerous "proofs of concept" out there that convincingly demonstrate that the environment is much more resilient than we give it credit for. The dust bowl of the American Midwest returned to tall- and short-grass prairie just 20 years after most experts in land management had written it off as a desolate, sterile region never to recover. It's no wonder they held this pessimistic view, when newsreel footage of that period shows toxic black clouds of topsoil about to engulf a whole town. Nonetheless, in the absence of any outside influence (i.e., farming), save for a few small restoration efforts on the part of the government, it quietly returned to its former ecological setting.

The entire northeastern portion of the United States was clear-cut at least three times in our history, and when it became apparent each time that farming could not succeed there, the land was abandoned and the trees obliged by returning in full force. An excellent example that has a solid scientific basis for describing what happens to a forest when it's been clear-cut, then left alone, can be found in the online records of the Hubbard Brook Ecosystem Study (http://www.hubbardbrook.org/), whose ecological setting is in northern New Hampshire. A summary of the study's initial results is instructive to those who remain skeptical that nature has the resources to recover from a catastrophic event like being clear-cut. When the study was initiated in 1967, an entire watershed's worth of forest was cut down, but the trees were left in place. The quality of the water draining the watershed was continuously monitored for dissolved minerals and organic compounds before, during, and after clear-cutting. Three years later, the quality of the water in the streams had returned to preexperimental condition. The trees took longer to grow back. At first, pioneer plants—shrubs and bushes—dominated the landscape. These were shade intolerant, and due to the continuous exposure of their seeds to sunlight, they rapidly germinated and filled in the wide-open spaces left by the downed forest. The pioneer plants grew rapidly, creating shade. Their roots helped hold the soil in place until the trees could take over that function, allowing the draining streams to return to their original high quality. The tree seeds in the soil that were now shielded from direct sunlight by the shrubs and bushes were stimulated to germinate. When the saplings had grown above the level of the shrubs and bushes (about five years) creating shade once again, the pioneer plants (shade-intolerants) died off, giving way to the regenerated forest. Within twenty years, the mixed boreal forest of Hubbard Brook watershed was once again standing tall and proud, without a single person lifting so much as a finger to help the process along. The revealed secret to reforestation that arose out of that seminal long-term ecological study (which is still ongoing, by the way) is the realization that all the seeds for the shrubs, bushes, and trees were in the ground awaiting their marching orders. The moment a disaster hits, nature goes into overdrive and ramps up the repair mechanisms. It is an amazing process to witness, regardless of which ecosystem is at risk, and reflects the cumulative knowledge of millions of years' worth of natural selection. In the end, nature always gets it right.

The demilitarized zone between North and South Korea, a ten-mile wide designated no-man's-land stretching from coast to coast, was declared off-limits in 1953 following the signing of the truce agreement in that same year. It is currently a verdant, peaceful wildlife reserve completely devoid of human activity. Once again, nature hung out its NO HELP WANTED sign and proceeded to regenerate itself. In another example, in 1986 Chernobyl, Ukraine, was the site of the world's worst nuclear power plant meltdown, resulting in huge amounts of radioactive fallout and contamination of the surrounding environment for miles in all directions. Amazingly, the wildlife did not obey the abundant NO TRESPASSING warning posters and slowly began to repopulate the region shortly after humans evacuated it. Today, it's a haven for wildlife. Almost all of the animals and plant species that were forced out of the region by farmers have returned. That is not to say that it's safe for humans to return to Chernobyl simply because it has the appearance of normalcy. The plants and animals are certainly paying a heavy price for their willingness to live in such a highly contaminated setting.

Costa Rica is a country rich in history that has enjoyed an unusually long period without internal strife. As a result, the landscape has filled up with farms, replacing the tropical rainforest with sugarcane, coffee, and forest plantations; a variety of short-cycle crops; and cattle. Despite heavy encroachment, approximately 47 percent of the country is still forested. It is understandable, then, that ecotourism has become a dominant player in that country's economy, with the Bosque Lluvioso rain forest as a primary destination for those seeking total emersion into a pristine tropical cloud forest. From 2000 to 2005, the country's cattle industry was forced to decrease in size due to the United States electing to buy cheaper livestock from other sources. This left large holdings of cleared land fallow. The Boyce Thompson Institute for Plant Research at Cornell University then began a restoration project that has resulted in the successful reforestation of a small portion of that damaged tropical environment, using state-of-the-art horticultural methods and seeds from local tree species. In addition, abandoned farmland in other parts of Costa Rica has also returned to its former ecological setting. According to a recent FAO report, *State of the World's Forests 2009*: "In most Central American countries, net forest loss declined from 2000 to 2005 in comparison with the previous decade, with Costa Rica achieving a net increase in forest area." Finally some good news about tropical forests!

Another restoration success story with more human input involved the recovery of a significant portion of the Samboja Lestari rain forest in the extreme eastern part of Borneo. In order to create a safe haven for more than one thousand baby orphan orangutans, Dr. Willie Smits, a wildlife biologist, through the help of local farmers, established a reserve by planting a series of native plants, including trees, on abandoned farmland. Before the plant life was restored, the farmland created by clearing forest had been parched and lifeless. In a short time, just three years after beginning the project, the pattern of weather returned to a tropical rain precipitation regime with daily episodes during the rainy season. The orangutans appear to be doing very well there, too.

Regrowth of forests provides several ecological services, the main one being sequestration of carbon in the form of cellulose, and the other restoration of biodiversity. However, how much carbon trees can remove from the atmosphere is somewhat controversial, and is still being investigated by a number of forestry researchers. It all seems to depend on what kind of trees (hardwood or conifer), how old and densely packed they are, and the latitude at which they are growing. One thing is certain, and that is that without trees, there would be few other ecological solutions for taking carbon dioxide out of the air. What's more, no other damaged ecological setting can be encouraged to repair itself simply by leaving it alone. For example, coral reefs are composed almost entirely of calcium carbonate and are very efficient at sequestering carbon. Unfortunately, they are on the decline due to rising sea surface temperatures. At the same time, the oceans have become saturated with CO_2, and instead of sequestering carbon at a higher and higher rate as atmospheric carbon dioxide continues to increase, the seawater is turning acidic as carbonic acid forms, leaving little else to help with balancing the atmospheric carbon budget.

So, for the sake of argument, let's assume we could somehow convince the states of Ohio, Indiana, Illinois, and Iowa to convert all their agricultural land back into hardwood forest, which most of it was prior to 1600. If it were allowed to regrow, that much hardwood forest would consume annually around 10 percent of U.S. emissions of carbon dioxide as it reached maturity (thirty to forty years). Of course, without an option for growing crops in a nontraditional fashion (controlled-environment agriculture), we would never even begin to contemplate what the effects of reforestation of the American heartland might have on climate change.

5. *No Use of Pesticides, Herbicides, or Fertilizers*

The vertical farm will employ state-of-the-art hydroponic and aeroponic technologies configured inside a secure building. The design of the building will take into account the need for keeping out unwanted diners, such as insects and microbial pathogens, that in an outdoor setting are able to more than satisfy their nutritional needs and raise hell throughout the world's agricultural landscape. The traditional farmer has to counteract this attack on crops with a variety of weapons, namely, pesticides and herbicides. In addition, fertilizers are essential for maximizing crop yields in nutrient-depleted soils. In contrast, the vertical farm will use pure water, into which a set of highly purified, carefully balanced nutrients will have been dissolved to satisfy the nutritional requirements of the plants. By adding additional nutrients that we also need, we will ensure that both the plants and the animals (us) will satisfy both parts of the equation. There will be no need to worry about contamination of our food with things like heavy metals, atrazine, diazinon, or human pathogens such as salmonella or *E. coli* 0157:H7. Enough has been already said about the adverse affects of these agrochemicals and contaminants on us and the environment without need for further exploration of this subject; suffice it to say that the data should convince any reader that it would be to our extreme advantage to avoid using them if we could. With the vertical-farm cultivation strategy, in which we have total control over everything, it would be possible to do so.

6. *Use of 70–95 Percent Less Water*

Today, traditional agriculture uses around 70 percent of all the available freshwater on earth, and in doing so pollutes it, rendering it unusable for those living downstream. In contrast, hydroponic, and more recently aeroponic agricultural technologies have revolutionized the way water is used to grow plants without the damaging side effects of agricultural runoff. When these two methods are employed in "closed loop," or self-contained, systems, a huge amount of water is conserved, up to 95 percent in some extreme cases. These two methods of cultivation are the NASA and the European Space Agency's answer for sustainable food production, and will enable astronauts to eventually produce food on the moon or Mars. Similarly, once vertical farming is perfected, food production on earth could take place anywhere. That is the long-term payoff behind the vertical farm project.

So how do these two related grow systems work? Contrary to popular belief, plants do not require soil, per se. What they use soil for is a solid base of operations into which they can spread their roots. In other words, the earth serves as a physical support system. That is why plants can be found all over the world, regardless of the kind of soil, as long as there is enough water and dissolved minerals, and a source of organic nitrogen. Provided the soil type does not adversely affect the plant by being too acidic or basic, then it's possible to get plants to grow nearly anywhere on the planet, even in the cracks of sidewalks or on the cliffs of mountains as bonsai plants. Plants can even colonize newly formed volcanic islands that have no soil at all. In fact, plants actually help create soil there by breaking down the large, rocky substrate into smaller and smaller particles through the growth of their roots until it comes to resemble a kind of primitive soil.

Hydroponics, developed in 1937 by Dr. William Frederick Gericke at the University of California, Davis, is the method of choice used routinely by nurseries to get seeds to germinate and sprout roots before they are transplanted into some form of potting soil. The avocado seed is probably the most familiar example of a plant that can grow to near maturity from a seed in a glass of tap water, with nothing more added except fresh air and sunlight. The reason it can achieve such advanced growth characteristics is that it has huge amount of stored nutrients in the seed itself. All it really needs to sprout—grow a stem and leaves—is water. Kids (and adults, too) love to watch it grow from a pale, tan, wrinkled, lifeless object into a brilliant green plant that, when transplanted into a pot of soillike material, continues to grow and often produces avocados if watered regularly. It is one of those small "miracles of life" that parents love to introduce to their children. Avocados are no different from all other plants in that they give their seeds the maximum chance for survival by storing lots of goodies in them. All seedlings will produce adult plants that can carry on with maintaining the species, given a reasonable climate regime. Life indoors will ensure that a "reasonable" environment is always there.

The essential elements that plants need are colored orange in the periodic table, and include an organic source of nitrogen. We, on the other hand, require at least seven additional elements (colored orange). So, when we design diets for hydroponically grown crops, we need to include all of the essential ones in the nutrient solution that bathes their roots. This is where I imagine large agrochemical companies could convert from synthesizing pesticide and herbicide manufacturing and become the suppliers of ultrapure, chemically defined diets for crops grown in vertical farms. It wouldn't take much in the way of economic incentives to convince that beleaguered industry to do the right thing and get on board the global green movement. After all, the executives and workers of those large corporations have families, too, and I am sure they all care about what happens to their own children and grandchildren over the course of the next fifty years.

Setting up a hydroponic facility is largely constrained by the kind of crop one wants to produce. The configuration is determined by the root system of the plant. The liquid portion of the operation is pumped slowly through a specially constructed pipe, usually made of a plastic such as polyvinyl chloride (or PVC), though it's not a requirement that plastic be used. Bamboo in various diameters could also serve the purpose quite well, and since it's one of the toughest natural materials we know of, bamboo would be ideally suited. Also, it's very easy to grow. There is no need to lock into any technological niche to begin with, however. Once the piping is set up, nutrients are dissolved into the water phase and circulated through the piping, all the while being electronically monitored for concentrations of each element and organic nitrogen. The result is uniform plant growth under optimal conditions. For a list of major hydroponic growers and the extensive variety of crops they have commercialized, see the Appendix.

Aeroponics, invented by Richard Stoner in 1982, takes hydroponics and "kicks it up a notch." Small nozzles located under the plants spray a nutrient-laden mist onto the roots, supplying them with everything they need. It is so conservative with respect to water use that it consumes about 70 percent less water than hydroponics, and will undoubtedly become a major player in the next phase of controlled-environment agriculture.

The question frequently arises: "Why don't the greenhouse tomatoes I have to buy in the winter taste as good as the ones I can grow in my own backyard?" I think in the "old days," the indoor varieties indeed were not up to the standards of soil-grown tomatoes. The reason is that when it became possible to grow commercial levels of produce indoors, growers strove to make their crops appear to be flawless to the consumer. They got exactly that: the perfect-*looking* tomato. However, one bite proved that you can never judge a tomato by its outward appearance; it was mushy and tasteless. The essence of a good-tasting tomato is hard to beat and easy to recognize, so the greenhouse industry began to look into why theirs didn't match up. By studying the outside conditions that produced tasty veggies (e.g., cold nights, warm days, or short periods of drought), they concluded that some stress was necessary in order to elicit flavonoids (complex organic molecules specific to plants). These molecules are the essence of why most vegetables have distinctive flavors and aromas. In addition, restricting the water a plant receives increases its sugar content, heightening the flavor even more. Today, many indoor growers have taken advantage of this information and now consistently produce the finest-tasting vegetables on the market. For example, EuroFresh Farms, located in Wilcox, Arizona, regularly wins blind-tasting competitions for its tomatoes. The greenhouse industries of the Netherlands and Mexico have also caught on to the new methods of how to produce tasty crops. Unfortunately, many small greenhouse operations still do not "get it." It's a constant tug-of-war between consumer and producer that will probably go on regardless of the fact that growers know what to do. For some, it's just too involved to care much.

7. *Greatly Reduced Food Miles*

There is something reassuring about the phrase "homegrown," whether it refers to a local football hero or the food we consume. Local is best because we know where it came from. The vertical farm is a neighborhood concept couched in futuristic terms, but with a homespun intent. The things we trust most are the things we can see for ourselves. Locally grown corn, tomatoes, or free-range chickens seem to taste better, also. We even brag about them to outsiders: Jersey tomatoes, Maine potatoes, Georgia peaches, and so forth. Michael Pollan stresses this concept almost to the breaking point in his epic foodie tome *The Omnivore's Dilemma*. The vertical farm will reside inside city limits and in doing so will create a local sustainable source of produce that will undoubtedly find its way into restaurants, school cafeterias, hospital bedside meals, prisons, and apartment complexes, as well as, of course, into the green markets. It will be freshly picked at the peak of ripeness, never frozen or even refrigerated. Its contents will be known down to the last atom of each element. It will all be sold at the end of the day. The amount of travel between the tomato and your plate will be measured in blocks, not miles. Ultimately, the United States will finally be able to reduce its carbon emissions. Remember, farming in the United States consumes some 20 percent of the fossil fuels used annually. Because food from the vertical farm needn't travel very far, its rate of spoilage will also be greatly reduced. No storage will mean less refrigeration and more fossil fuel saved. The urban waste scene, now swarming with vermin, may be greatly reduced, as well. Is there a downside? Well, urban properties are prohibitively expensive compared to even the most expensive farmland. I will deal with this issue in another chapter. For now, be reassured that even in New York or Los Angeles, there is affordable real estate where the presence of a vertical farm would transform the most neglected of neighborhoods into a wellspring of urban rebirth and vigor. I am convinced by what I have learned over the last few years of traveling and speaking to city councils, mayors, city planners, and ministers of agriculture that if vertical farms become firmly established inside the city, it will be because these officials saw virtue in them and created incentives to bring them on board.

8. *More Control of Food Safety and Security*

The vertical farm, regardless of configuration, should be constructed in such a way as to exclude most known plant arthropod pests and microbial pathogens by using the same principals that are applied to the design and construction of intensive care units for hospitals. Barrier medicine has been successfully developed and deployed over the last hundred years, ever since we first became aware of pathogenic microbes and their characteristics. This approach will enable the vertical farm to operate pest- and pathogen-free for the great majority of the time. Prevention is key. Reacting to an invading plant disease or insect pest is expensive, time-consuming, and inefficient. Vertical farming would become impractical if it had to shut down every six weeks or so to address an outbreak of something like whitefly, a common visitor to most open greenhouses. Once inside, this tiny pest can reduce profitability in several weeks, and must be dealt with before production can be resumed. Rice blast, wheat rust, and a number of other plant pathogens must be controlled by exclusion, not with antifungal agents. Positive-pressure buildings with filtered air supplies, secure locks, and workers who must change their clothes before entering will guarantee that the vertical farm is a safe and secure place to raise our crops. Andrew Carnegie offered the world his philosophy on protecting the things we treasure most: "The fool saith in his heart: Do not put all your eggs in one basket. But the wise man exhorteth thus: Put all your eggs in one basket; then watch the basket." I am almost certain that he was referring to the money he had amassed and stashed away in some bank. Yet bank robberies do occur from time to time. Nowadays, high-tech safes and a cadre of well-armed guards stand between the Butch Cassidys and Willie Suttons of the world and our hard-earned deposits. Similarly, by centralizing agriculture into the urban landscape, it will be necessary to ensure that no one can easily sabotage the operation. Secure entryways and badge-only admittance of approved workers will be essential in this regard. On a traditional farm, there are plenty of weak points from a security standpoint. I expect that the concept of vertical farming will become applied to a variety of situations—restaurants, schools, hospitals, and apartment complexes, for example. Decentralization of food production, once it is situated inside the urban landscape, will go a long way to thwarting any terrorist activities. Remember: Outside, we control nothing, while inside, we get to control everything. The choice is ours. I choose indoors every time.

Workers in vertical farms will have to be screened for certain groups of parasitic infections that could be spread by fecal contamination, the way the City of New York used to screen food handlers before certifying them for working in restaurants. Economic constraints prohibit such an approach today in New York, but I would argue that detecting organisms such as geohelminths and salmonella, then treating those few infected workers before they go on the job, could eliminate the potential for spreading food-borne illnesses once vertical farms become the primary source of produce in the city.

9. *New Employment Opportunities*

The advent of vertical farms will create numerous new opportunities at many levels. Municipalities will use vertical farms to rehabilitate urban spaces once considered too degraded to serve as commercial properties. Areas in which vertical farms become situated will, in turn, attract new development, making the urban food desert a thing of the past. A 2009 survey of urban farmers in New York City, San Francisco, and Portland, Oregon, revealed that they began their interest in agriculture after committing to a life in the city. They wanted both to live an urban lifestyle and to raise some of their own food. Most were self-taught, but as farming inside tall buildings catches on, it will generate a new set of careers: managers, indoor controlled-agriculture specialists, waste-to-energy specialists, and farmworkers for the nursery, planting, monitoring, harvesting, sorting, and selling. New industries associated with developments in hydroponic and aeroponic grow systems will become the new

"Silicon Valley" specialty industries, along with sophisticated electronics corporations manufacturing instruments for everything from germinating seeds to monitoring nutrient delivery systems to harvesting crops. Groups of specialized workers will undoubtedly form organizations that will offer their services to the varied vertical-farm industries, such as the owners of small versions situated on the roofs of apartment complexes, restaurants, and hospitals. Value-added manufacturing establishments will also spring up next to the larger vertical farms to process fish, shrimp, shellfish, and poultry, and to take advantage of the year-round availability of fresh, vine-ripened fruits and vegetables. Grain production is not out of the question, either, and mills could become located in cities. The year-round production of hops and barley in some vertical farms may even ignite a renewed interest in city breweries. Wine grapes grown in vertical farms could make vintage wines as commonplace as bottled water.

Within the last five years, less developed countries have experienced countless crop failures, caused mostly by adverse weather events such as floods and droughts. No matter where the meeting or what the announced purpose, gatherings of world leaders invariably include speeches about the need to develop new strategies for overcoming hunger and poverty, and they all cite agriculture as the number-one priority in starting to turn things around. With the failure

Courtesy of Peek & Cloppenburg KG (www.peek-cloppenburg.de)

A CURVED, HIGHLY TRANSPARENT BUILDING DESIGN FOR THE VERTICAL FARM Designing to take full advantage of natural sunlight is highly desirable. Those versions of the vertical farm able to do so will be highly energy-efficient and, perhaps, even fully independent of the energy grid. Orienting the vertical farm to take maximum advantage of the daily progression of the Sun across the horizon is one option for achieving this goal.

In this instance, the farm might take on the form of a crescent-shaped transparent building, such as the one depicted here. Getting the sunlight into the back of the VF will necessitate the use of specially designed parabolic mirrors. Commercial versions of these reflective devices are currently available. The use of fiber optics to bring sunlight to individual plants might also be another option for some vertical farms.

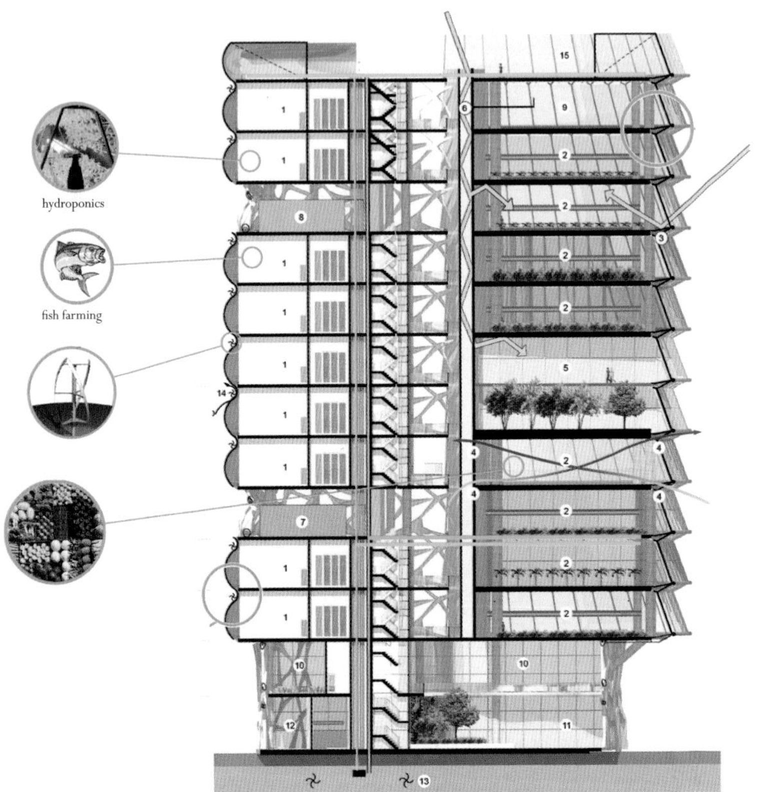

1. HYDROPONICS, AQUAPONICS, PROCESSING OR ACCOMMODATION (MIXTURE)
2. CROP SECTIONS (LARGER CROP TYPE FARMING)
3. REFLECTIVE EDGE OR LIGHT SHELF
4. STRATEGICALLY PLACED VENTS PROVIDE MULTIPLE VENTILATION SCENARIOS (TO FURTHER CATER FOR EACH PLANTING VARIETY)
5. ORCHARD SECTION (MORE INTENSIVE FARMING)
6. LIGHT TUBE—MAXIMIZING THE NATURAL LIGHT
7. PLANT LEVEL—LOCATION IS FLEXIBLE
8. WATER STORAGE LEVEL
9. RESTAURANT
10. CAFÉ/RESTAURANT
11. ENTRY
12. STORAGE
13. WATER TURBINES (DEPENDENT UPON LOCATION)
14. WIND TURBINES
15. ROOFTOP FARMING

This "stackable" vertical farm was designed in Australia by ODESIGN.

Oliver Foster / O Design (www.odesign.com.au)

(*left*) Fish farming is ideal for combination with indoor farming as the nutrient rich water developed from the fish can then be used for the plants/crops. A 5600 liter tank can provide 800 fish.

(*right*) The omega garden carousel is a hydroponic power house.

- each one uses just 14m^2 of floor area
- rotation develops oils in the plants which enable faster growth rates and better taste
- uses LED lights (18.2kw/day)
- uses 99% less water than tradional farming

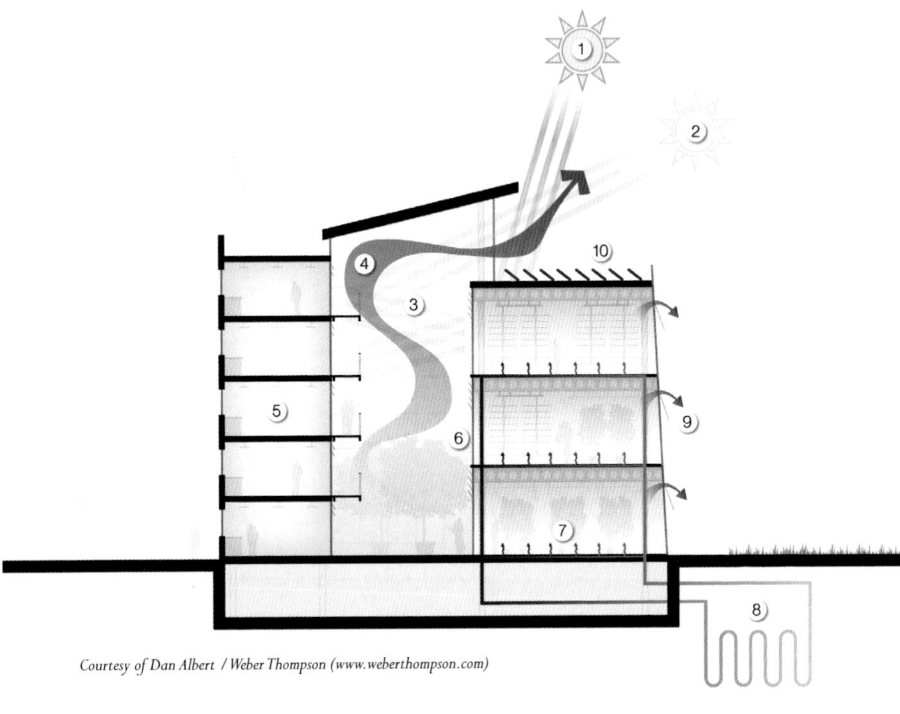

Courtesy of Dan Albert / Weber Thompson (www.weberthompson.com)

THE ENERGY SYSTEM

1. summer sun
2. winter sun
3. reflected light
4. thermal stack
5. north side—cool thermal mass
6. warm air vented from greenhouse
7. radiant floor
8. ground source loop
9. operable vents
10. photovoltaic panels

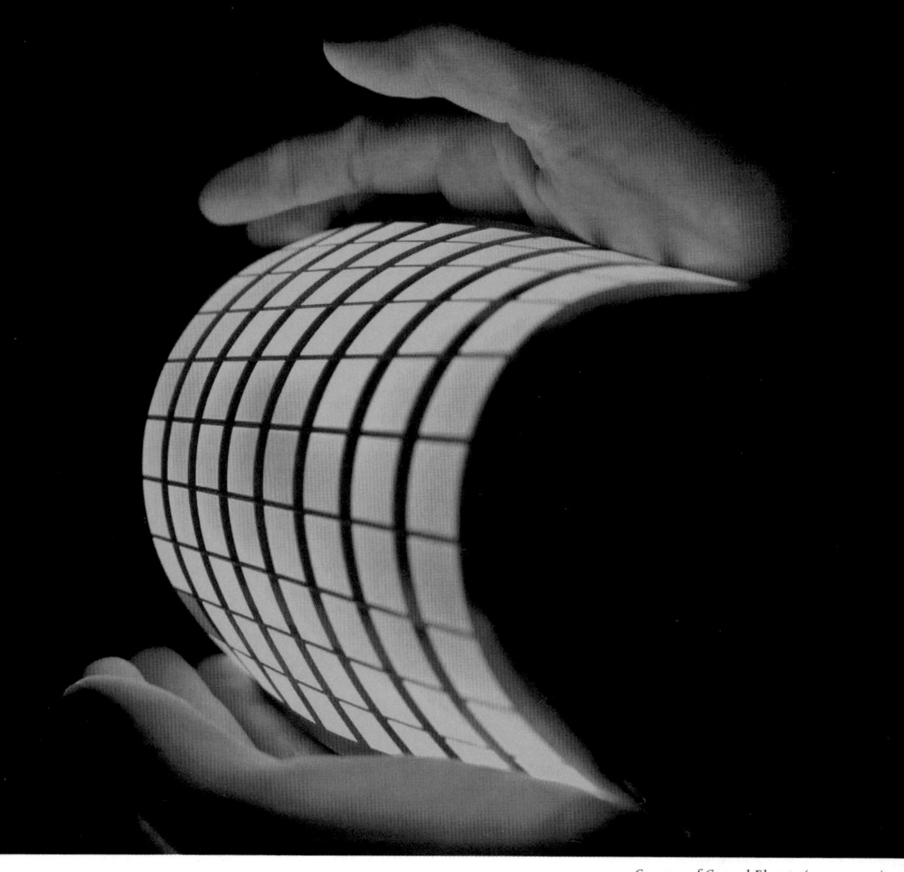

Courtesy of General Electric (www.ge.com)

ORGANO-LIGHT EMITTING DIODE Getting enough of the right kind of light into the vertical farm might require the application of new technologies that are just now maturing. Ordinary tungsten-generated light is composed mostly of wavelengths that are useless to plants, making them highly inefficient for use in greenhouses. The same is true for fluorescent lighting. Light-emitting diodes (LED) currently exist and have been engineered to give off narrower wavelengths of light, roughly 400 and 700 nanometers. All higher plants carry out photosynthesis using two kinds of chlorophyll, a and b, and each chlorophyll compound absorbs light at either one of these two wavelengths.

While LED lighting is good, the use of light-emitting organic compounds spread out on thin films of plastic (OLED) allows for an even greater control of the wavelengths of light. Because OLED reduces the wavelength to exactly what is required by higher plants, much less energy is consumed by the lighting fixture. In addition, because OLED lighting consists of a flexible plastic film, it can be physically manipulated to allow it to be configured around individual plants.

Courtesy of Jung Min Nam / JN_Studio, Thesis Project at GSD Harvard University, 2009 / Advisor: Prof. Ingeborg Rocker

NATURAL VENTILATION

SMALL-SCALE: Cross Ventilation Local passive ventilation through cross ventilation.

TOWER, HOUSING, AND OFFICE LEVEL Each housing and office unit has cross ventilation. Semipublic farming area and greenhouse activate the overall ventilation along the central chimney.

WATER PURIFICATION SYSTEM Open space required for the aerobic bio reactor enhances the air drift by the stack effect.

PRIMARY FARMING AND ATRIUM Farming area has cross ventilation as well as stack-effect ventilation along its connected inside space. Large vent at the public lounge helps the overall stack-effect ventilation.

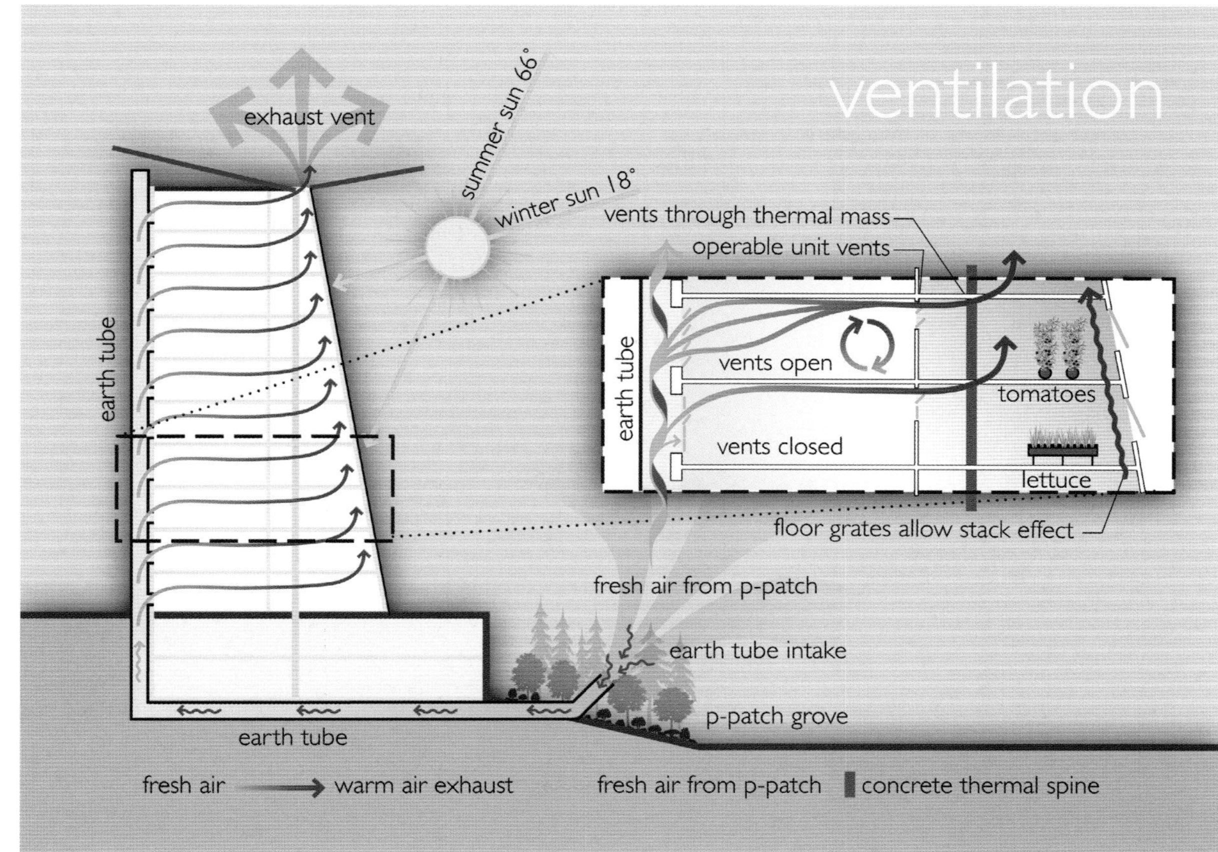

Courtesy of Dan Albert / Weber Thompson

Courtesy of Eurofresh Farms

EUROFRESH FARMS, WILCOX, ARIZONA This 318-acre highly successful commercial greenhouse complex is located in the middle of a barren portion of the Arizona desert. It drills deep into the rock for its water and recycles all of it. The only moisture that leaves the farm is in the form of tomatoes and cucumbers. Its produce regularly wins blind tasting contests, providing ample evidence for the validity of indoor farming as a viable alternative to the current traditional farming situation. By "stacking" high-tech indoor farms, such as the ones designed by Eurofresh Farms, on top of each other, we can create vertical farms and locate them near city centers, providing urban dwellers with convenient, fresh, healthy produce year round.

Courtesy of Weber Thompson Architects, Seattle (www.weberthompson.com)

This is a proposal done by Weber Thompson Architects in Seattle for a prototype vertical farm complete with hydroponic grow areas, grey water remediation, research facilities, living quarters, retail space, and a learning center for children. A vertical farm of only a few stories high would be ideal to determine how best to integrate all the necessary technologies, in order to achieve maximum efficiency.

Courtesy of Romses Architects (www.romsesarchitects.com)

(*above and right*) FOOD, ENERGY, AND WATER HARVESTING
The concept of "harvest" is explored in this project through the vertical farming of vegetables, herbs, fruits, fish, and egg-laying chickens. In addition, renewable energy will be harvested via green building design elements harnessing geothermal, wind, and solar power. The buildings have photovoltaic glazing and incorporate small- and large-scale wind turbines to turn the structure into solar and wind-farm infrastructure. In addition, vertical farming potentially adds energy back to the grid via methane generation from composting nonedible parts of plants and animals. Furthermore, a large rainwater cistern terminates the top of the "Harvest Tower," providing onsite irrigation for the numerous indoor and outdoor crops and roof gardens.

Vertical Gardens Wind Turbines Solar Panels Hydroponic Carousel

Courtesy of Romses Architects (www.romsesarchitects.com)

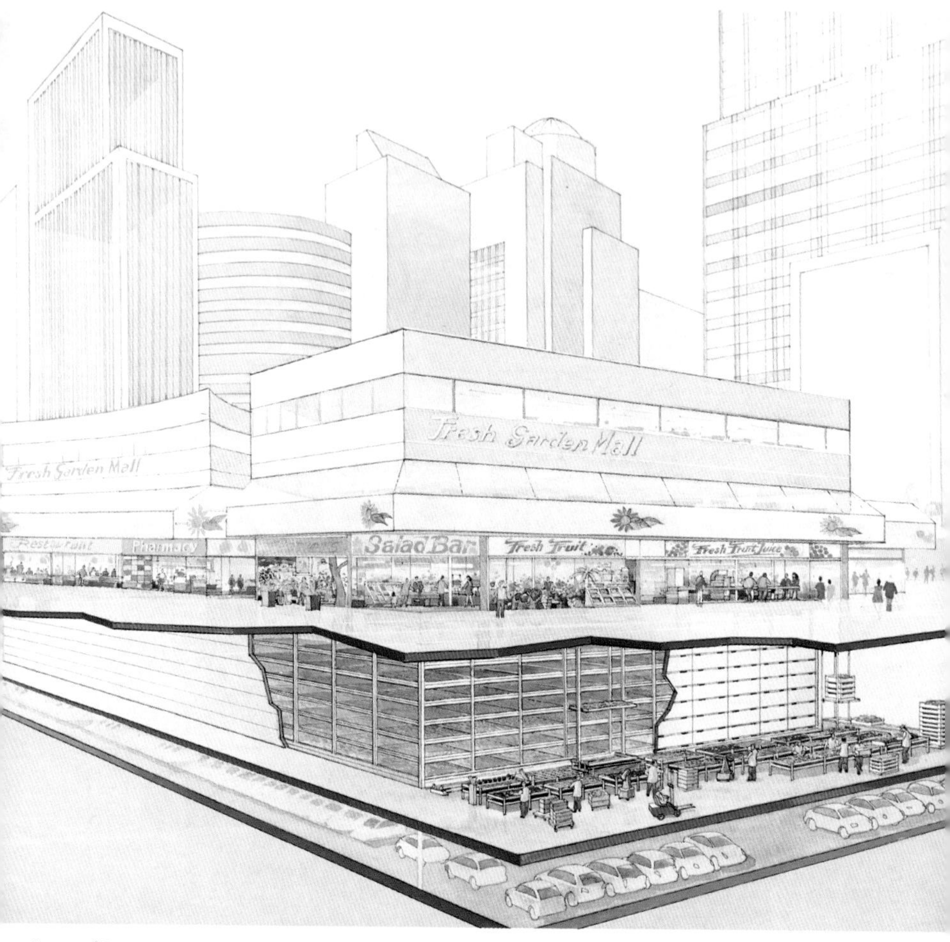

Courtesy of Gertjan Meeuws, WTIc / 's-Hertogenbosch, Netherlands

PlantLab is under construction in Den Bosch, the Netherlands. It is privately owned and operated. One of its unique features is that it is located underground. It will produce leafy green vegetables, peppers, cucumbers, and other varieties of produce. Like the two previously pictured vertical farms, PlantLab employs LED lighting and is highly automated.

Courtesy of Mr. Min, supervisor of the vertical farm / Korea Department of Rural Agriculture

The vertical farm owned and operated by the Republic of Korea opened in March of 2011. The building is three stories tall, features LED lighting, and all the hydroponic grow systems are highly automated. It produces leafy green vegetables.

Courtesy of Kojiro Ishii, Director of Strategic Initiatives, COO / Green Green Earth, Inc. / Beverly Hills, California

Nuvege is a four-story vertical farm about the size of a large 747 airplane hangar located in Kyoto, Japan. It is privately owned and operated and produces leafy green vegetables. It employs state-of-the-art hydroponics and LED lighting. Most of the production schemes are automated.

Courtesy of Vincent Callebaut Architectures (www.vincent.callebaut.org)

THE DRAGONFLY TOWER Two marinas accommodate two huge aquaculture ponds. This intriguing design was developed by Vincent Callebaut.

Courtesy of Dr. Dickson Despommier

THE APPLE STORE AT FIFTH AVENUE IN NEW YORK CITY This iconic structure is the entryway into the Apple Store located in the basement of the building. It is made of high tensile strength glass and it's totally transparent. Imagine a vertical farm some five times the volume of this structure and several stories in height. Made of ethylene tetra-fluoro-ethylene (ETFE), not glass, the vertical farm of the author's dreams could be lightweight, totally transparent, and fully insulated with a double skin of ETFE, making it both attractive and plant-friendly.

Courtesy of Scientific American

FUTURE FARMING

of farming in already-stressed-out regions of the globe, urbanization has increased at a rate disproportionate to the birth rate. Most of the migrants are composed of farmers and their families. What better groups to have work in vertical farms than those who already know how to farm? The future looks bright for the creation of new jobs for the new industry of vertical farming.

10. *Purification of Grey Water to Drinking Water*

Each day, every city produces huge amounts of grey water derived from black water by the removal of solids. New York City, as mentioned earlier, produces and then throws away an astounding 1 billion gallons of grey water a day. It is the responsibility of each community to discard their waste in ways that do no harm to the environment. This is the basic plan for all municipal sanitation codes, but waste management has proven difficult to execute and very expensive to sustain. Billions of dollars are spent each year in the United States figuring out new ways of disposing of liquid municipal waste. It is time to realize that these products of our own metabolism have intrinsic value and must be turned into recyclables, and the energy in the solids extracted by some form of high-tech incineration scheme. For recovering the water, plants have the answer. They are called living machines for that reason. In brief, plants obtain their nutrients by pumping water up through their roots, through their leaves, and then out into the atmosphere. This process, referred to as transpiration, allows them to take up nutrients in the form of elements and organic nitrogen. The elements and nitrogen stay inside the plant and become incorporated into new tissues of the growing parts of the organism, while the water is continually transpired through tiny pores in the leaves called stomata. Remediation of grey water could easily be accomplished by taking advantage of this basic plant activity inside vertical farms constructed solely for that purpose. In this case, the plants would not end up as someone's salad; that would be too risky from a public-health standpoint. Capturing the water of transpiration can be accomplished after the plants have pumped the grey water through their tissues, purifying it before releasing pure H_2O into the enclosed atmosphere of the vertical farm. Dehumidification of the indoor air is all that would be needed to get back the water we produced by eating and drinking. It might require a second round of transpiration to completely purify the grey water, but this may not even prove necessary once an experimental system is up and running. In conventional greenhouses, the humid air is expelled outside by window fans. If New York City were to establish a water-recovery system based on the above strategy, and sold the water for as much as two cents per gallon, a year's worth of NYC grey water turned into drinking water would be valued at $720 million. Quite a windfall over the long term, considering the expense the city now goes to in order to treat and discard it. As Senator Everett Dirksen of Illinois once said, "A billion here, a billion there, and pretty soon you're talking about real money."

11. *Animal Feed from Postharvest Plant Material*

Conserving energy will be a concern if the vertical farm uses a significant amount of electricity to grow crops. In this case, incinerating the postharvest portion of crops would be a viable strategy for energy recovery. In other situations that do not require this, leftover plant material could be consumed as animal feed, depending on the crop.

CHAPTER 6

THE VERTICAL FARM: FORM AND FUNCTION

Change is the essence of life.
Be willing to surrender what you are
for what you could become.

—ANONYMOUS

In its most complete configuration, the vertical farm will consist of a complex of buildings constructed in close proximity to one another. They will include a building for growing food; offices for management; a separate control center for monitoring the overall running of the facility; a nursery for selecting and germinating seeds; a quality-control laboratory to monitor food safety, document the nutritional status of each crop, and monitor for plant diseases; a building for the vertical farm workforce; an eco-education/tourist center for the general public; a green market; and eventually a restaurant. Aquaculture and poultry will be housed in adjacent but separate buildings with no physical connection to the vertical-farm building to ensure safety for the plants. I will limit my discussion here to the buildings devoted to plant-based agriculture and leave the fascinating subject of indoor aquaculture and the raising of fowl for a later discourse on the eco-city. A vertical farm that might accommodate four-legged animals would not be practical, nor would it be humane. That having been said, ultimately, I will not be the one to control the choice of what goes into the vertical farms. In many societies, pigs would be a natural addition to the farm. I doubt, however, that anyone would contemplate raising goats, sheep, or cattle inside one.

"Form follows function" is the defining principal for life on earth. The environment selects examples at every level that best emulate this "golden rule" of nature, right down to the shape and mode of action of the molecules we're made of. The wings of a bird or a bat have much in common, obviously because they both have the ability to fly. The long, muscular legs of an ostrich are built for running, mostly from predators, while the eight-jointed, spindly legs of a spider enable it to nimbly manipulate itself about on its silken, sticky fly trap. It's the same when we design for ourselves. Louis Sullivan, the acknowledged dean of American architecture, stated the concept in powerful, sweeping syntax, as Moses himself might have done were he to have somehow given himself permission to add an eleventh commandment governing the aesthetics of the human condition: "It is the pervading law of all things organic, of all things physical and metaphysical, of all things human and all things super-human, of all true manifestations of the head, of the heart, of the soul, that the life is recognizable in its expression, that form ever follows function. This is the law."

When planning the vertical farm, architects and engineers must be driven by this critical concept, since the vertical farm will be built to satisfy the needs of the crops and not necessarily ours. Failure to do good science to bring it all together will all but guarantee a repeat of the Biosphere 2 fiasco. Despite the fact that no vertical farms currently exist, some general notions can be established ahead of time that would apply to any version of one. The concept of farming in tall buildings inside the city limits has slowly fermented into a unique broth of ideas over the last three or four years. This "reaction mixture" is composed of numerous inputs besides those that occurred in my classroom: midnight conversations over half-empty glasses of wine (notice I didn't say exactly how many times it had been totally empty); exhausting question-and-answer sessions and extended conversations following presentations to numerous professional groups; as well as academic lectures at schools of architecture, engineering, and planning. I have also been asked to present my thoughts on vertical farming to branches of the USDA and the United States Agency for International Development. I can now, for the first time since this project was conceived in 2000, envision what a vertical farm might look like. Through the many conversations I've had with all of those who share my passion for urban farming, I have further refined my ideas as to how to proceed. Amazingly, I have received little in the way of negative-toned feedback encouraging me to redirect my efforts, or heard that I am out-and-out crazy and that the vertical farm will never "fly." A large number of images of vertical farm designs (shown in the signature section of the book and on the Web site www.verticalfarm.com) preempted these conversations. The vast majority of these wonderful, and often futuristic, designs were unsolicited. They arrived to me unannounced, and, I might add, very much appreciated for their ability to stimulate thought without necessarily addressing the functionality issues I am about to discuss. Therefore, these colorful, creative objects of imagination are more in the way of "eye candy" and "what ifs" than practical applications.

What follows is based on the construction of an experimental prototype vertical farm and associated essential buildings (nursery, laboratory, control center), first referred to in chapter 4. The farm itself would be modest in height, perhaps five stories tall and maybe ⅛th of a city block in footprint. The interior space would be highly flexible, allowing the expert team of controlled indoor-agriculture scientists the maximum freedom to configure and reconfigure the conditions the crops will be subjected to. They would work closely with local communities of consumers, a team of material scientists and structural engineers, and a state-of-the-art construction facility for the selection of crops and the custom manufacture of hydroponic and aeroponic equipment. Undoubtedly, many new modes of growing would be invented under these ideal research conditions. All of the experimentation would be of the applied variety for the sole purpose of eventually being able to make the vertical farm as productive and profitable as possible.

THE VERTICAL FARM BUILDING

Most crops have a fairly broad range of tolerances with respect to temperature and humidity. This will enable the indoor farmer to mix and match a wide variety of plants and to grow them in the same room if desired, as long as their root systems are held at the optimal temperature for each species. In designing for the tenants, success goes to the farmer who can best manage temperature, humidity, and security. This is the "holy trinity" of indoor controlled-environment agriculture.

There are four major themes that designers and engineers must include in any version of a vertical farm:

1. Capture sunlight and disperse it evenly among the crops.

2. Capture passive energy for supplying a reliable source of electricity.
3. Employ good barrier design for plant protection.
4. Maximize the amount of space devoted to growing crops.

The materials employed in the construction of the building will be dictated by the needs of the plants and secondarily by the needs of those who work inside the vertical farm. That is not to say that the environmental conditions will become intolerable for humans. Quite the contrary, plants and people go well together, so the temperature and humidity profiles maintained inside the building should allow for a very pleasant work environment as well as favoring maximum crop yields.

1. CAPTURE SUNLIGHT AND DISPERSE IT EVENLY AMONG THE CROPS

We Are on the Same Wavelength

Designing a large, secure home for plants requires intimate knowledge of what a plant needs and how it all works together to allow for maximum growth. Those readers already familiar with the inner workings of a plant can feel free to skip this brief discourse on photosynthesis and go to the next section. For those not familiar with basic plant physiology, the following information may clear up a lot of confusion as to why the family potted plants are not doing so well. Hopefully, it will help you turn your black thumb into a green one.

Plants are fundamentally different from animals in that plants require water, a few elements including carbon dioxide, a source of organic nitrogen, and sunlight to grow. The key here is sunlight. The main plant ingredient that allows all of this, and the part that we are missing, is a microscopic organelle called the chloroplast, and there are hundreds of thousands of them in each leaf. It is a complex structure with its own genome. Chlorophyll is the familiar green-colored compound residing inside the chloroplast that captures sunlight in the form of photons. Chlorophyll then does its magic, converting photons into chemical energy that is used to link together the carbons found in carbon dioxide, forming sugar and other plant-specific products (e.g., cellulose). In the process of photosynthesis, plants discard the oxygen portion of carbon dioxide into the atmosphere, supplying all animals with one of the essential elements they require to carry out their own lives. When we eat plants, we derive the sugar (and, of course, other nutrients) from their tissues. We combine the oxygen we breathe in with the carbons of the sugar molecule, one carbon and two oxygen atoms at a time. This produces carbon dioxide and chemical energy in the form of adenosine triphosphate. We then use that chemical energy to construct our own tissues, and breathe out carbon dioxide as a waste product. Plants take up carbon dioxide and begin the cycle all over again. It's a remarkable mutually linked association. It should be noted here that the vertical farm, in addition to producing our food, will also sequester huge amounts of carbon dioxide from the atmosphere and, most important, produce lots and lots of oxygen. So every time a worker inside the vertical farm breathes out, you will almost be able to hear the plants say, "Thank you."

There are two major forms of chlorophyll, chlorophyll a and chlorophyll b. Both absorb light in two distinct wavelengths of the visible spectrum, blue and red (roughly 400 to 700 nanometers). Plants contain lots of other phytopigments, too—carotenoids, for example—but they play only a minor role in photosynthesis. The take-home lesson is that not all the energy in sunlight is needed to grow any crop to its maximum yield. We can take advantage of this fact by creating lighting exclusively for the plants. Light-emitting diodes (LEDs) have already been specifically engineered to do that, resulting in a significant savings in energy costs. In contrast, conventional lightbulbs emit 95 percent of their energy as heat (very inefficient, to say the least) and the rest as a broader spectrum of light, most of which is useless to the plant. On the near horizon are organo-light-emitting diodes (OLEDs) made of thin, flexible plastics. These contain stable organic compounds that allow for even more narrow spectra of light to be produced, saving more energy and money, while still giving plants exactly what they need. In addition, OLEDs will permit the design of lights that could be made into any configuration, placing the light source at the optimal distance from the plant, regardless of the plant's shape. They could even be wrapped around each growing plant, offering the ultimate in custom, energy-efficient lighting for our food crops.

Here Comes the Sun

In areas of the world that already enjoy the gift of abundant sunlight—for example, the Middle East, Australia, the American Southwest, many parts of sub-Saharan Africa—using the sun as the only source of energy to grow crops would be entirely feasible and highly recommended. Photovoltaics could easily supply the energy needed to run any electrical equipment, while sunlight would supply all the energy needed to grow the crops. Orienting the footprint of the vertical farm with a north-south exposure will allow designs to capture maximum amounts of light. Narrow, long, and low—three to five stories high and perhaps as long as a half mile—might be the paradigm for designing in those sun-drenched environments, where land adjacent to urban centers is cheap and available. Buildings with deeper interiors could take advantage of newly developed, specially constructed composite plastic parabolic mirrors such as those produced by Sunlight Direct, situated outside and behind the building to first concentrate, then direct sunlight to the interior sections, while the front of the building remained exposed to the maximum amount of light. Fiber optics leading from the collecting mirrors outside to the inside of the building for each floor could further assist in the distribution of energy to the plants. Together, these two approaches should allow for any reasonable design, regardless of its ultimate shape. A crescent-shaped structure would present a uniform surface to the sun as it progressed across the horizon each day, making this design the most efficient for using passive sunlight. In that case, no further lighting would be necessary, unless twenty-four-hour grow cycles for some crops were an option.

The Visible Farm

If sunlight is the main source of energy to grow the crops, then the vertical farm should be made as transparent as possible. The designer/architect has many choices of transparent material to choose from. Glass is cheap to manufacture and durable, albeit a bit on the fragile side and heavy. Since the late 1950s, glass-and-steel combinations have led the way into the future of the skyscraper. Some of the best-known applications of glass-and-steel construction are two New York City icons of early modern architecture, the Lever House, designed by Gordon Bunshaft of Skidmore, Owings, and Merrill and built in 1952, and the Seagram Building, designed by Ludwig Mies van der Rohe, and completed in 1958. A current trend in modern building design advocates for total transparency; the Apple Store on Fifth Avenue in New York City is an excellent example of what we can expect to see more of over the next twenty years. It is now even possible to create an all-glass structure without any metal at all in the building by using special adhesives. The caveat here is that, as of this writing, the new glues used to attach sheets of glass together have only been tested for a year's worth of wear. Insulating an all-glass building is a big problem, and employing double-glazing to provide energy-savings adds huge amounts of weight and expense to the equation.

One solution is to abandon glass altogether in favor of high-tech plastics that are much lighter in weight and more durable. Recycling transparent plastics (bottles, etc.) into clear panels used for windows and modular construction of eco-friendly structures has spawned a new industry for building materials. A leading exponent of this approach is

KieranTimberlake Architects in Philadelphia. The only difficulty is that most commonly available plastics yellow over time due to excessive exposure to UVB radiation from unfiltered sunlight, excluding more and more of the wavelengths of light that plants need to grow efficiently. One newer product, referred to simply as ETFE, or ethylene tetrafluoroethylene, is a fluoropolymer plastic with many advantageous properties, including the fact that it is a "self-cleaning" material, due to its electrostatic charge. It is very lightweight, only 2 percent the weight of glass of a similar thickness; is as transparent as water, allowing in all wavelengths of light; and has a high tensile strength. Most important, it does not yellow when exposed to sunlight for long periods of time, making it far superior to any other lightweight clear carbon-based polymer material on the market. ETFE has been applied to such iconic buildings as the Beijing Olympics swimming venue, the Water Cube, designed and constructed by the China State Construction Engineering Corporation jointly with Australia's PTW Architects and Ove Arup Pty Ltd; and the Eden Project in the south of England, designed by architect Nicholas Grimshaw and the engineering firm Anthony Hunt and Associates. Both of these high-profile projects made extensive use of ETFE. These structures are now several years old and remain in good condition. Creating a double or even triple skin of ETFE for the outer glazing ensures good insulation quality and reduces the need to use large air-conditioning units or much in the way of heating. Maintaining a positive pressure inside ETFE also creates cushions of insulation. I would opt for constructing my prototype out of aluminum framing and large panels of pressurized ETFE to allow in the maximum amount of sunlight to the crops inside. A positive pressure will also allow for maximum safety and security designs, building into the vertical farm double-lock entry systems. In this configuration, invasion of the growing zones by microbial pathogens and insect pests would be greatly reduced.

2. CAPTURE PASSIVE ENERGY FOR SUPPLYING A RELIABLE SOURCE OF ELECTRICITY

Earth, Wind, and Fire

While sunlight will be the main source of energy to grow crops in regions that have more than two hundred days of sunlight, many other regions would be left out if this were the only viable way of proceeding. Scandinavia, Iceland, most of Russia, Canada, and Alaska in the United States in the Northern Hemisphere; and Chile, Argentina, and New Zealand in the Southern Hemisphere would all need to tap into an alternative energy supply to remain independent of the municipal energy grid. Fortunately, there are many choices, and in some of the countries mentioned above, generous supplies exist. These include geothermal, tidal, and wind energies.

The Fire Down Below

Geothermal sources are abundant in the United States, Iceland, Italy, Germany, Turkey, France, the Netherlands, Lithuania, New Zealand, Mexico, El Salvador, Nicaragua, Costa Rica, Russia, the Philippines, Indonesia, the People's Republic of China, Japan, and Saint Kitts and Nevis. It comes in several varieties: naturally occurring surface vents of steam or water, such as those found inside Yellowstone National Park, and an abundant gradient of heat from molten magma that lies at or just below the surface in places such as Italy, Iceland, and Hawaii. A third source, the geothermal heat pump, is proving to be very useful in modern construction practices and is not limited to any specific geological formation. This method can be used to either cool or heat a building. The Department of Energy describes the process thusly: "Using the Earth as a heat source/sink, a series of pipes, commonly called a 'loop,' is buried in the ground near the building to be conditioned. The loop can be buried either vertically or horizontally. It circulates a fluid (water, or a mixture of water and antifreeze) that absorbs heat from, or relinquishes heat to, the surrounding soil, depending on whether the ambient air is colder or warmer than the soil." Installing current versions of these devices for individual homes has proven to be expensive because they are relatively new, but like all other things manufactured, as demand goes up, prices will go down. Once installed, they will eventually pay off themselves by the savings on fossil fuel, and will always be good for the environment. Applying geothermal heat pump technology wherever we can to the first prototype vertical farm will ensure that it will never be a negative asset to the community with respect to energy consumption.

Blowin' in the Wind

The new "oil," according to energy magnate T. Boone Pickens, is wind power. He has identified the entire American Midwest as the next Saudi Arabia. According to his own calculations, the United States could save some 20 percent of its total energy budget spent on fossil fuel use by converting the wind-power capabilities that that geographic region enjoys into electricity. The area's flatness plays a major role in the abundance of wind power there. Other areas of high wind production are the world's coastal regions; the farther north or south one goes toward either pole, the higher in strength and more reliable the wind is due to the rotation of the planet. So as long as we continue to go around and around, there will always be wind for the taking. Today, many countries have tapped into that resource and in doing so have significantly reduced their energy bills and improved the quality of the atmosphere at the same time. These countries include Canada, the United States, Germany, Spain, the Netherlands, Denmark, Sweden, India, and China.

First-generation wind turbines were based on the same principles of the old-fashioned Dutch windmill. While efficient at capturing power from the rotation of large propellers, they were not without their own unique set of unintended consequences, such as a high number of bird deaths. In addition, the blades would eventually slow down due to the friction created by the gradual accumulation of insects on their exposed surfaces. The early model generators did not yield as much electricity as theoretically possible, but they were a beginning that would eventually spawn an entire clean energy industry. The new generation of wind turbines are even larger in size than the original models and are equipped with more efficient generators. As a result, these turbines turn slower, capture more energy per turn, allow birds the luxury of seeing them so they can easily avoid collisions, and do not need as much cleaning since insects do not accumulate as rapidly as before. In fact, older-model wind turbines that have received these new generators have improved their efficiency by 10 to 25 percent. All in all, it's been a victory for the ingenuity of wind-energy engineers, who successfully worked through each and every problem. Only one problem remains, and that has to do with aesthetics, not functionality. The old NIMBY cry can still be heard in many locations when the wind turbine issue comes up before the town council meeting.

Other wind-capturing devices have completely broken out of the old windmill mold, assuming radically different configurations. One successful design is a horizontal double-propeller-shaped wind turbine that resembles an old-fashioned hand-driven lawn mower. These new devices are attractive, efficient, and require less wind speed to operate than conventional wind turbines. Furthermore, their turning mechanisms are quiet, do not place unwanted stress on the buildings, and are easy to add on to existing structures without major reconstruction efforts. By combining wind turbines with photovoltaics, a solar/wind capture strategy can be effective in generating energy both day and night.

Burn, Baby, Burn

The vertical farm will produce food, but it will also produce a significant amount of inedible plant and animal by-products (i.e., waste). In a traditional farming operation, or even with the vast majority of low-tech greenhouses, the postharvest leftovers are typically plowed under to partially supply next year's crops with a jump-start of nutrients, or discarded into the trash bin. Organic material, regardless of what form it takes, is a valuable resource that begs for use in any energy-recapture system. It is good to keep in mind the fact that the word "waste" does not appear anywhere in the ecosystem's dictionary. It's all part of the same natural loop of energy recovery aiding in the regeneration of life. If the vertical farm is to behave like an ecosystem, then the roots, stems, and leaves of crops, and the entrails of fowl and fish, all need to find their way back onto the energy grid. Incineration is the most practical way to proceed. Composting was for several years considered by many waste-to-energy experts to be a viable option for handling municipal waste streams. For small-scale situations like backyard lawn clippings and leftovers from dinner, a family living in the suburbs would still do well to compost them and employ the products of worm metabolism—so-called worm casts—as a fertilizer in gardens. Upon further analysis, though, it became apparent that the efficiency of composting was too low for anything commercial in scope: Giving 80 to 90 percent of the energy contained in rotting organic waste to the microbes in exchange for a 10 percent "return on investment" in the form of methane is a no-win technology. In addition, there is residue to contend with after the anaerobic digestion process of composting comes to an end. Much higher efficiencies of energy generation can now be achieved by incinerating biomass with devices that produce minimal levels of pollutants while giving off heat for the steam generation that runs the turbines that make electricity. Most of Europe now employs some form of incineration to process its solid and liquid municipal waste streams back into valuable kilowatts of electricity. With the advent of plasma arc gasification (PAG) devices, any solid material can be reduced to its elements in a matter of seconds. PAG uses an electrical current to create a high-energy plasma, the fourth state of matter. The plasma forms when two electrical arcs unite in the center of the combustion chamber. The device requires that material first be reduced in particle size to accommodate the narrow spray nozzle that introduces it into the device. In the case of liquid municipal waste, it is only necessary to dilute it in order to achieve the proper viscosity. The intense heat of the plasma arc (approximately 4,000 to 7,000°C) exceeds that found on the surface of the sun. Pyrolysis is the process that vaporizes all compounds that pass in front of the plasma arc back into their elements.

The heat released is used to make steam and generate electricity. Plasma gasification of a single ton of solid municipal waste would generate approximately 800 kilowatt-hours of electricity that could then be added to the grid or used directly by the vertical farm. The process itself uses around six times less energy than it generates. The other advantage is that at the end of the day there is no residue to deal with. Recovering energy from the inedible parts of the harvest (stems, leaves, roots, etc.) makes the vertical farm energy-efficient and opens the way for entire cities to behave similarly.

3. EMPLOY GOOD BARRIER DESIGN FOR PLANT PROTECTION

Better Safe Than Sorry

Food security and safety issues have to be dealt with as two sides of the same coin and are the primary concerns of the vertical-farm management team. Both have risen to the top of the list that forms the basis for a national food safety program coauthored by the USDA and the Department of Homeland Security. Outdoor farming represents an open-ended, no-holds-barred battle to the death between the crops we plant and those things that would consume them before they reach our plates. Outdoor control strategies, for the most part, consist of programs designed to limit the spread of a given insect pest or disease by the application of pesticides or herbicides or by culling only the affected portion of the crop, all the way up to destroying the entire year's planting. Inside, things will be quite different and much more controllable. The exclusion of unwanted visitors by applying positive pressure to the building housing the crops and nursery buildings will be the first step. Designing double-lock-entry doorways will allow for an additional level of protection against insects and microbes. Requiring all personnel to change into sterilized, disposable safety uniforms, shoe, and hair coverings, and to shower before changing clothes, will minimize the risk of crop loss due to "hitchhikers" on items such as shoes. Because the vertical farm will not need fertilizers, the risk from contaminating plants with human pathogens will be all but eliminated. This must be coupled with an initial—followed by an annual—routine series of laboratory tests for all vertical-farm workers designed to detect carriers of salmonella, giardia, cyclospora, and the like. (See the table on pages 200–201.) Animal pathogens such as *E. coli* 0157:H7 will have no chance of contaminating the produce of the vertical farm, since no large herbivores will be housed anywhere in or near the complex. If a security breach does occur in the vertical farm resulting in contamination of the crop, then destroying the entire crop can be dealt with the very next day. Once the security leak has been identified and fixed, the vertical farm could resume full production within a reasonable length of time. On the outside, the farmer must wait until the following spring to begin again, and often with the same disastrous results. Frustration is indeed the mantra of the outdoor farmer.

FAECALLY TRANSMITTED INFECTIONS

VIRUSES

Rotavirus

Hepatitis A

Hepatitis E

Polio virus

Norovirus

Adenovirus

Astrovirus

BACTERIA

Salmonella typhi

Shigella

Vibrio cholerae

*E. coli strain 0157:*H7

Camplyobacter jujuni

Helicobacter pylori

Clostridium difficile

FUNGAL

PROTOZOA

Entameba histolytica

Giardia lamblia

Blastocystis hominus

Cryptosporidium parvum

Cyclospora cayetanensis

Endolimax nana

HELMINTHS

Hookworm

Trichuris trichiura

Ascaris lumbricoides

Fasciola hepatica

Schistosoma mansoni

Schistosoma japonicum

Heterophyes heterophyes

Ophistorchus viviarini

Paragonimus westermani

Clonorchis sinensis

One additional idea for early detection of invading microbes should be mentioned, although it is not quite off the drawing board yet. Imagine the deployment of a genetically engineered nonedible plant species (arabidopsis comes to mind as a prime candidate for this kind of job) to monitor for the presence of a variety of plant pathogens using a molecular biological approach. The plant could be modified with a reporter molecule such as fluorescent green protein that has been stitched together with a snippet of DNA from the offending microbe. The "canary in the coal miner's cage" plants would be interspersed among the crops. Simply by turning off the lights at night and turning on the UV light, one could tell if any individual plant has encountered the pathogen, since it would glow green. Then culling that area would be simple and could save an entire floor's worth of crop plants. Just a thought.

4. MAXIMIZE THE AMOUNT OF SPACE DEVOTED TO GROWING CROPS

How to configure each floor of the vertical farm will depend solely on the crops selected. As the hydroponics industry matures, many different growing modalities will arise to meet the challenge of maximizing the yield for all the crops inside. Today, there is enough experience with hydroponic technologies to offer the bare-bones essentials of what is available and what we might expect to see over the next few years given the current rate of progress. Many plants are "comfortable" growing in the traditional hydroponic piping complete with holes. The plants are spaced at a similar distance apart, as one might see on a traditional soil-based farm. Tomatoes, lettuce, spinach, radicchio, green beans, peppers, zucchini, cucumbers, cantaloupes, and many others fall into that category. Grains are best grown in sheets of inert material similar in consistency to a spongy fiberglass air conditioner filter. One new approach for growing aeroponic grains uses a Dacron-based clothlike sheet as a matrix for the roots that the seeds are spread onto and then germinated. Nutrients are sprayed on the sheet from below. Lighting then becomes the biggest challenge, especially when growing several layers of the same crop in the grow room. The hydro-stacker employs a form of drip irrigation for growing potted plants such as strawberries, eggplant, and avocado. Even corn can be grown hydroponically in large tubs, usually with six plants per tub. Each plant yields around three ears per plant, and a crop matures every eight to ten weeks, allowing for at least five crops per year. If one wanted to, the system for growing could be reconfigured onto a conveyer belt–like hydroponic system. Producing corn indoors in the vertical farm actually could become quite profitable and allow for the reclaiming of many acres of land at the same time.

Piping can be made from a wide variety of materials, but most of today's hydroponic/aeroponic equipment is made of some variety of plastic. Polyvinyl chloride (PVC) plastic is readily available and is the most commonly employed material for constructing hydroponic grow systems. Leaching of toxic phthalates from PVC into the nutrient solution is a concern and can be minimized by first treating the PVC with a dilute sulfide solution. This treatment cross-links the plastic, trapping the phthalates permanently, and eliminates any health risk that might be associated with leaching. Some people would object to the use of any plastic, however, due to environmental considerations, mostly what happens to the plastic after it's discarded. The accumulation of all varieties of plastics in landfills and aquatic environments such as the open ocean is absolutely appalling and unacceptable. This kind of pollution is now totally preventable by instituting some form of environmentally friendly incineration. I would agree that burning any plastic product in a conventional (i.e., low-tech) incinerator produces noxious compounds that have serious health risks associated with them, but this situation can easily be avoided by vaporizing PVC and other plastic polymer products in a plasma arc gasification system. If we really get clever, perhaps we can find a use for various diameters of hollow bamboo and forget about plastics altogether. It is one of the fastest-growing plants, it's strong, it does not rot when kept wet, and it can be harvested to match any desired diameter of piping.

Undoubtedly, as vertical farming becomes perfected over the next few years, innovation at the level of configuring the grow space for all major crops will win the day. Until then, I can only offer this meager list of suggested modes to address the most important issue of maximizing the yield.

CRADLE TO CROP

Once the construction of the vertical farm complex is complete, the next step will be to purchase and receive starting materials (i.e., seeds). Where to get seeds for any crop is not a trivial issue, since there are many varieties of each cultivar to choose from. Fortunately, there are organizations specializing in supplying seeds (e.g., Siegers Seed Company, Florida Foundation Seed Producers, Inc., Good Seed Company, Neseed) that have served the commercial outdoor and hydroponics communities well, and their names can be found in publications that cover the greenhouse industry. One publication that I have found particularly helpful and informative is *Practical Hydroponics & Greenhouses*. The Food and Agriculture Organization is another superb resource for this kind of information, as is the USDA. Some seed producers have genetically engineered their own crops, particularly corn and soybeans, to resist things such as drought and herbicides. Monsanto is one of those corporations and has a strict policy about the use and production of its seeds (see the film *Food, Inc*). In my view, restricting the use of seeds for any purpose limits their application to large commercial outdoor establishments where adverse weather and invasion from weeds is a constant problem. The whole rationale behind the vertical farm is to avoid these problems to begin with.

The seeds must first be surface decontaminated, then sent to the diagnostic laboratory for testing for the presence of microbial pathogens that might be commandeering them as their Trojan horse. Once certified disease-free, the seeds will be sent to the nursery for quality-control testing and germination. The nursery will be a separate facility from receiving, as the nursery is the first chance any pathogens might have of directly contaminating the inside of the vertical farm. Since the germinated seedlings will eventually have to enter the vertical farm, security must also be maintained in this building. The nursery and vertical farm will most likely be connected by a maximum-security pressurized lock system. Seeds will first be evaluated for their ability to germinate and grow. All plant crops will originate in the nursery as germinated seedlings, and once germinated they will be tested again for any pathogens that might have slipped through the first screening. The infant crops will be transferred to the vertical farm and situated into their hydroponic/aeroponic environment. All crops will be constantly monitored by remote sensing systems for growth and nutrient conditions. Much of the work in the nursery will be labor-intensive, creating many new job opportunities for those with a green thumb.

COOL, CLEAR WATER

Water for the vertical farm will be used in hydroponic and aeroponic growing situations, and for the workers for their showers and drinking water. It could come from several sources, depending upon the geographic location and the ability of the urban community to access grey water for reuse. The highest-quality water should be used whenever possible. Usually this means drilling a well or obtaining water from a river, lake, or reservoir that is then filtered before being applied to crops. The obvious advantage of controlled-environment agriculture is the fact that it is a "closed loop" system, thereby allowing for the capture of water vapor derived from transpiration by employing dehumidification devices on each floor. This represents a highly efficient system of water use for farming compared to the traditional mode of outdoor soil-based agricultural irrigation schemes. In the closed-loop system, as mentioned earlier, hydroponics uses some 70 percent less water than conventional farming, while aeroponics uses 70 percent less water than hydroponics. In either case, it is a substantial improvement, and for areas of the world in which water is already in short supply, switching to vertical farming is the only reasonable approach to allow more water to be made available for drinking purposes. As also pointed out earlier, no runoff occurs in the vertical-farm model. If implemented on a large scale, vertical farming would have the possibility of eliminating ocean pollution from agricultural runoff. This has to happen if the productivity of the world's estuaries is to be restored.

WHAT'S FOR DINNER?

The question often arises as to which crops can be grown indoors. The answer is a surprising: "Anything you want." All one has to do for proof is visit any well-maintained botanical garden, for example, Kew Royal Botanic Gardens in London, the Bronx Botanical Gardens in New York, or the Missouri Botanical Garden in Saint Louis. Nearly every kind of exotic plant can be found inside these wonderful facilities. The New York Botanical Gardens, for instance, houses and cares for the world's largest (and, it might be noted, smelliest) flower, *Rafflesia arnoldii,* which is quite rare and grows in splendid isolation in the dense tropical forests of Indonesia. Fortunately, it rarely blooms, or the whole place would always stink of rotting flesh. If horticulturists can manage to grow that oddity, then anything is possible. As for the edible plants, one needs to consider several things before choosing which ones to grow. First, there are economic considerations. Is it worth the effort? Can farmers sell out the crop each and every time it's harvested, and at a profit? Can it be presold to commercial buyers? If so, all the better if profit is the main driving force behind that particular vertical farm. So far, a few popular vegetables have been grown successfully for profit. These include tomatoes, lettuce, spinach, zucchini, green peppers, and green beans. The strawberry has also been a star for the indoor farmer. Hardly any of these crops would qualify for addressing the needs of a hungry world, however. Essential crops such as wheat, barley, millet, rice, and potatoes would be more appropriate. The answer as to whether or not we can grow these crops indoors is still yes. All of them have been grown hydroponically. If their successful production means that a country that had to import nearly all its produce can now, within its own borders, supply its own population with an essential healthy diet, then profitability takes a backseat to need. Government-sponsored food programs may become the determining economic factor in the form of incentives and subsidies that enable the vertical farm to survive and even thrive, producing crops that in a free-flowing market economy would ordinarily fail to generate enough income to make them worthwhile.

An online resource that is very helpful in all aspects of hydroponic farming is the National Sustainable Agriculture Information Service (http://www.attra.org/). As one can see, the choices of crops that could be produced in the vertical farm far exceeds any consumer demand, regardless of the cuisine, with the exception, of course, of the meats derived from livestock.

DANGLING MODIFIERS

Undoubtedly there are other topics that one should take into consideration when designing, building, and then operating the vertical farm, but I think most of the important ones that deal with its "nuts and bolts" and general systems have now been laid down, albeit in a highly truncated executive summary form. More detail regarding any of these areas would require sitting down with a team of experts in a wide variety of fields related to architecture, agronomy, engineering, and the like and brainstorming until a blueprint for the prototype vertical farm emerged. I don't claim to have enough expertise or insight to offer any higher level of description of any of the topics covered in this chapter, so

I am afraid we will all have to wait until such a team is assembled and given their marching orders to find out what happens next. There is one missing area I can address, however, and that is who will work in the vertical farms, and who will benefit most from their creation.

CHAPTER 7

THE VERTICAL FARM: SOCIAL BENEFITS

Our only security is our ability to change.

—JOHN LILLY

The vertical farm is the keystone enterprise for establishing an urban-based ecosystem. Without food production, no city can emulate the virtues of a functional, intact ecosystem: Bioproductivity is key for both. It is the defining mechanism for energy management for all living organisms. Yet, if the city can supply itself with at least 50–80 percent of its agricultural needs, then lots of other sustainable activities become achievable, allowing its citizens to capture and reuse the energy of their own metabolic by-products and reclaim the grey water. Establishing vertical farming on a large scale would be the start to a complete remake of urban behavior centered around the concept of doing no harm to the environment. Ultimately, it is about creating a healthier lifestyle for anyone living anywhere in the city, making the built environment an ideal place to raise children, and about improving the overall environment of the planet.

UNCONTROLLED GROWTH IS THE PHILOSOPHY OF THE CANCER CELL

Architects and city planners are fond of referring to cities as "living organisms"; cities have a soul, a life of their own, a unique personality. They lean toward describing metropolises in ethereal, romantic, abstract terms, waxing creatively and often misusing basic English in the process. I have heard some even go so far as to state that the city is a new form of "organic" superbeing, evolving out of the collective energy of all those who have had a hand in shaping its past, present, and future. If that is true, then despite all the nice descriptors, hype, metaphors, and genuine cultural advantages of living in one, in reality the city has assumed the role of a monstrous parasite when viewed from an ecological perspective. It sucks up prodigious quantities of the earth's raw materials, gulping down the nutritious parts in a single, noisy, pollution-producing swallow, then spews, sprays, flings out waste of all kinds onto its own doorstep and well beyond. Seen through the eyes of the natural world, a modern city is a twisted amalgam of concrete, steel, and glass crammed full of two-legged, very aggressive life forms; a place to be avoided at all costs, save, of course, for those animals—rats, mice, sparrows, pigeons, squirrels, cockroaches—opportunistic enough to take advantage of the absence of large predator species. In fact, there is a highly successful urban ecosystem, sans humans, filled with secretive denizens of the night that consume piles of unused biomass in the form of curbside garbage. It's estimated, for example, that New York City supports some ten rats per person. That's 80 million rodents, and what do you suppose they eat? Not garbage; it's too risky. They consume a diet rich in cockroaches, insects that gladly take chances to gain access to the restaurant wastes generated each day by the more than twenty-eight thousand food establishments within the city limits. Surely we can find a better use for that organic refuse than to enable and energize the roach/rat ecosystem.

Regardless of location, the city has grown helter-skelter, and its insatiable appetite and out-of-control metabolism produces nothing more useful than lethal bubbles of heat and contaminated air and water laced with the by-products of its mechanized infrastructure. "Metropolis" has become synonymous with "consumption." None of this negative behavior was planned, yet urbanization over the last hundred years turns out to be a thousand times more destructive than all wars put together, both in the scope of the planetary damage it has created, the number of human deaths caused by unhealthy living conditions, and its penchant for continuing to cause even more disruption of the natural world on an ever-increasing scale, as new methods for construction are established. Godzilla is a mere toddler's hand puppet compared to the way the city itself has risen up into the surrounding landscape and crushed it flat with its big foot of progress. Atlanta, Georgia, is a case study that NASA features on its Web site, which shows how that city grew over a twenty-year period as documented by LandSat satellite imagery. Older case studies include those governed by the Aztecs, Mayans, Romans, and other failed cultures, eloquently described by Jared Diamond in his book *Collapse.* Granted, we have come a long way compared to the ancient Romans, who routinely discarded their garbage right out their windows onto the streets of the Eternal City, but in many cases, our "garbage" still comes back to haunt us just the same. Landfills, brown fields, rodent-infested abandoned city lots, nonpoint source runoff, coal-burning power stations, refineries of all kinds, not to mention all forms of vehicular traffic, add insult to an already injured biosphere. Encroachment is what we do, even though every time we extend the urban boundary, there is usually a heavy health risk associated with it. This is especially true in the tropics. For example, the building of the Transamazonian highway resulted in construction workers becoming infected with several new varieties of disfiguring dermal leishmaniasis. Yellow fever became more prevalent along its course, and countless species of wildlife became extinct due to extensive deforestation.

Today's urban health hazards come in many forms: air pollution, water pollution, noise pollution. In the United States, for example, there are more than a few cities that have unhealthy levels of fine particulates and surface ozone. Some of the worst places to raise children or to grow old are Pittsburgh, Los Angeles, Newark–New York, Houston, Dallas, and Baltimore–Washington, D.C. Mexico City is one of the world's worst cities for air pollution. Its air is often so contaminated with fumes from auto, bus, and truck exhaust, as well as propane stoves, that it can actually be seen; eyes water, noses run, and hospital admissions from asthma attacks soar. No city is completely immune from this kind of environmental insult. Drinking water is another essential resource and varies greatly in quality, depending upon the city in question. In 2001 the Natural Resources Defense Council ranked the quality of drinking water for many of America's largest cities. Among the poorest were Boston, Albuquerque, Phoenix, San Francisco, and Fresno.

MY DOG'S BIGGER THAN YOUR DOG

There is a new iteration of architectural imperialism: up. Building it bigger does not mean building it better, but bigger and bigger the buildings get. I was in Dubai in March 2008 when the Burj Dubai became the tallest constructed object on earth. Precisely what purpose that building will serve is not clear, even to its developers, who are still deciding on how many floors it should have. But it has, in the meantime, consumed huge quantities of steel and concrete, and at least one human life. Something has to change, or . . . well, I think I've spent enough time on this depressing subject. We need to look into the future and find ways to reinvent ourselves in the image of a hybrid techno-biosphere entity, using the best of what is available to reconfigure things in favor of eco-friendly solutions to the global dilemmas of climate change and urbanization. A few communities worthy of mention are planning in the right direction, closing loops on things like waste management and energy consumption: Vauban, Germany; Curitiba, Brazil; Växjö, Sweden; Toronto, Canada; Kampala, Uganda. None of them has yet incorporated urban food production into its sustainability plans.

FOR THE GREATER GOOD

What good will the vertical farm bring to the built environment? Philosophically speaking, defining a social benefit can be tricky. It is often predicated on the principal of whom the benefit does the most for. The one I favor is this: A social benefit should empower the vast majority of the citizens of any given community. Such things as fluoride in drinking water to help prevent cavities, public health services associated with food inspection, universal health care (hm . . .), social security benefits, public transportation and schools, community hospitals, and the like all fall into the category of "for the greater good." If this all sounds too ideal for the real world, well then, so be it. Aiming high never got anyone tarred and feathered, although I can think of several brave politicians—Adlai Stevenson and Bill Bradley, for example—who didn't get elected president because they embraced this philosophy. I am not naive enough to believe that the vertical farm will exist mainly for the benefit of the world's underserved communities, although I certainly wish that this could be so; on the contrary, there is the real possibility that the first couple of vertical farms might end up benefiting the few (commercial growers) and not the many. My students have expressed to me more than once that they have great concern that the idea, however well intentioned, may end up as a financial success only for those with the money and power to make it happen. Unfortunately, I am afraid I will not be able to do much about this, since the idea is already out in the public domain. As with the first televisions, cell phones, automobiles, and so forth, the wealthy will seem to have ready access to them due to their initial high cost and scarcity, while the not-so-well-off will have to wait a bit longer for their versions to enter the marketplace. The poor will have to wait longer yet, but since sometimes they also get what everyone else now has—witness the explosion in sales (and low cost) of the modern cell phone—there is hope that they will have easy access to the vertical farm, too.

Since the first vertical farms are likely to be prototypes, hence experimental in nature, I don't think large numbers of people will benefit immediately from them, except for those research teams working in them. Governments might become directly involved in financing and developing the concept to feed the majority of its citizens. Countries such as Norway, Sweden, the Netherlands, Denmark, and Australia have great need for vertical farms and are also economically empowered to fund such long-term projects. In those instances, the vast majority of their citizens would benefit sooner, as the research would be funded by contracts and grants. A good example of what happens to a good idea that keeps getting better is the way we invented the method for sequencing our own genome. In the beginning, things were time consuming and labor intensive. The right technical innovations had not yet appeared on the scene. Then, as the scientific community demanded faster and more accurate results, the inventor part of our brain got to work and came up with newer, faster, cheaper machines that saved the day. Today, some twenty years down the road, we are contemplating sequencing every species' DNA.

THINK GLOBALLY, ACT GLOBALLY

Benefits from establishing a vertical-farming industry at the global level would result in creating the next green revolution. Such an industry would bring a measure of stability to unstable regions of the world, for example the Middle East, that now bicker and fight over scarce water resources and lack diversity in their diets due to limited farmland and a harsh desert climate. Blights of plant diseases (e.g., wheat rust, rice blast) would be greatly reduced, and crops lost to locust invasions, mostly in West Africa, would be a thing of the past. It would mean we finally had control over our own destiny with regard to where our daily bread came from. For countries that routinely run out of food, it would mean no more starvation or malnutrition, and for those more fortunate ones it would mean less farming outdoors and more land available to return to nature. This is probably the number-one environmental reason for creating vertical farms, and I have given a more complete description of the positive consequences in chapter 5.

The implications for a sustainable food supply without further damage to the environment are obvious. The general health status of the world's children would rapidly improve. Reduction in the infant mortality rates due to starvation or from diarrheal diseases transmitted by fecally contaminated water and food would be greatly reduced. These two problems are the most serious public health issues that will be addressed by the vertical farm in areas where staggeringly high rates of infant mortality are the driving forces in population increase. People naturally have more children because few survive to adulthood. Other pressing health problems that would also experience a dramatic reduction in incidence and prevalence would include all forms of malnutrition, especially obesity and type 2 diabetes.

One possibly negative aspect of the globalization of vertical farming would relate to international commerce agreements dealing with agricultural produce. These trade agreements might need to be reestablished based on products not amenable to this mode of production (i.e., cattle).

THINK GLOBALLY, ACT MORE LOCALLY

Speculating about the possible social benefits of vertical farming requires that it be restricted in terms of cultures, since each country is unique, altering the "take" on what a vertical-farm industry would mean for it. I leave this daunting task for someone else, as I have no intention of going through each and every one of the 192 member states recognized by the United Nations and giving my impressions of what the vertical farm would look like and how it might function there. That would be downright silly and pretentious. However, qualifying vertical farming's social-benefits package based on political systems is one way to present the variety of ways in which the concept might manifest itself, and one which I feel more comfortable discussing.

I'll begin with the issue of governmental stability. In a peaceful environment, almost anything is possible. As of August 16, 2009, however, some thirty countries were at war or experiencing widespread civil unrest, and would be excluded from the "luxury" of establishing some form of the vertical farm. Not that it could not happen, but the likelihood of such countries maintaining vertical farms would be remote, at best; they would most likely become a target for the opposition. Nine of these unfortunate places are in sub-Saharan Africa. It is a gross understatement to point out that these countries in particular are the ones that most need rescuing from hunger, starvation, and the myriad infectious diseases associated with strife. In chapter 8, about alternate uses for the vertical farm, I will discuss the possibility of creating a MASH-like unit for portable vertical farms, specifically designed to address the issue of starvation and malnutrition among those who are displaced by war, civil unrest, and natural disasters.

Among the world's stable governments, the democracies of the G-20 nations will most likely take the path of least resistance and leave the development of vertical farms up to the private sector, with encouragement in the form of government-sponsored research programs. Socialist countries such as Sweden, Finland, Norway, and Iceland will most likely take a more active role in their establishment. Most countries throughout the Middle East have great need for the concept of vertical farming. Some of them have the financial ability, due to the fact that they produce huge amounts of oil and natural gas and are controlled by a few leaders (i.e., a monarchy), to assemble and fund the teams needed to do so at the snap of a finger. I anticipate that the Middle East will become a world center for vertical farming, and sooner rather than later.

The World Bank could also become a major player in financially helping to establish vertical farms in countries that cannot afford them, but that have great need. Many of these countries are in sub-Saharan Africa (e.g., Niger, Chad, Mali, Malawi, Senegal, Côte d'Ivoir, Central African Republic, Uganda, Zambia, Botswana, United Republic of Tanzania, and Kenya). Private foundations and nongovernmental agencies could also contribute to their development. Finally, private investors could step up to the plate.

EVERY DAY GETS A LITTLE BETTER

As the concept matures through applied research, and versions of the vertical farm begin to insinuate into the free market place, more and more citizens will reap the benefits of working in or living near one; of this I am certain. When the advantages of vertical farming become known and accepted by the majority of an urban community, then consumer pressure will ensure that more vertical farms will be constructed, bringing new markets to their town. What will a vertical-farming industry mean for any community living near one? Actually, it's quite simple: jobs, jobs, and more jobs. Jobs at all levels. It is an easy thing to imagine who might work in the vertical farm, as well as who might benefit most from having them in their communities. The number and kind of employment opportunities relate to the robustness of the vertical-farm concept. Early versions will contain the essentials, but little else. Nonetheless, skilled professionals—managers, developers, architects, engineers, planners, agronomists, waste-to-energy personnel, sales personnel, educators, security personnel, laboratory personnel (microbiologists, molecular biologists, technicians, and supervisors), as well as a large unskilled labor force whose job descriptions will vary from overseeing the harvesting and the delivering of produce from the vertical farm to local green markets, to managing the waste streams generated by harvesting, are the major categories of employment opportunities that will be available once the vertical farm complex is built. Each one of these jobs has an essential role to play in the running of the complex, and most important, all of these new jobs will be music to the ears of local government and community leaders.

The vertical-farm building, being as transparent as air and housing green plants from floor to ceiling, irrespective of its final shape, will be a radical departure from the standard architectural model of glass-and-steel construction. Because of this essential feature, and the fact that it will actually do something besides look good, it will be an instant hit with architectural journals, city planners, design schools, eco-urbanists, city farmers, futurists, schoolkids, the popular press, and the media, placing it squarely in the face of the entire planet. Some buildings that have already attracted this sort of attention include the Apple Store at Fifth Avenue and Fifty-ninth street in New York City, and the ING headquarters building in Amsterdam. Both of these new iconic structures turned millions of heads the moment they imposed themselves into their cityscapes. Showcasing the virtues of urban high-rise agriculture with the vertical farm and demonstrating its essential contribution to sustainability with an adjacent eco-learning center will guarantee a steady stream of tourists, who will flock to get a firsthand look at the infant future city. This in turn will generate much-welcomed tourist dollars. Urban properties, long abandoned and decaying, will become prime real estate for vertical farms, since brown fields will not pose a problem due to the fact that everything will be self-contained. This will make available huge tracts of marginalized land, such as obsolete industrial parks. Inner-city abandoned lots could be easily converted to vertical-farm complexes, not only adding much-needed revenue to the city coffers, but also eliminating "food deserts." The inner cities of most American metropolises have few if any high-end grocers like Whole Foods. City officials give a variety of excuses when asked why this is so, but the most common ones relate to high property-insurance rates and lack of political pressure from those living there to improve the situation. In many instances, the dominant ethnicities of the inhabitants of inner cities are minorities; Asian, Hispanic, and African American. Bringing the vertical farm to those areas will be like a breath of fresh air, boosting morale, especially among individuals who have lived all their lives in those communities and who have become intractably cynical due to political marginalization. Because vertical farms will by their very nature be things of beauty, challenging any modern expression of the new architecture for "best in show," neighborhoods will take great pride in welcoming them into their midst as stunning and nurturing parts of the local scene, as highly desired as a new park. Populations of unwanted residents (rats, mice, cockroaches) will plummet due to waste-management schemes that will have been "piggybacked" onto new, more efficient, less polluting ones established expressly for the vertical farm. New cuisines may even arise with more varied produce available 24/7. With vertical farms scattered throughout the urban landscape, city life will start to reflect the essentials of ecological process, producing food and recycling all waste. When this finally occurs, the promise of a sustainable, healthy future will be well within citizens' reach. This, then, is the ultimate social benefit: to live long and prosper.

CHAPTER 8

THE VERTICAL FARM: ALTERNATE USES

Change starts when someone sees the next step.

—WILLIAM DRAYTON, SR.

Food production is only one example of how the vertical farm could contribute in a positive way to urban life. Plants have the capacity to help out in many other ways, too. They are often referred to as "living machines," a term first coined by John Todd, the venerable systems ecologist best known for his work on environmental remediation. In 1969 Todd founded the New Alchemy Institute and began to design microecosystems, self-contained plant communities inside greenhouses. Using them as small-scale prototypes for larger real-world projects, he revolutionized how we approach solving environmental cleanups using specific plant species to sequester things such as heavy metals, pesticides, herbicides, and other toxic chemicals from contaminated wetlands, estuaries, and lakes. Today, there are many examples of how plants have been put to work making our built environment a greener and safer place to live in. One outstanding case in point is the rest stop along Route 89, just north of White River Junction in Vermont. Constructed in 2005, this eco-friendly complex remediates black water into safe, usable water for growing decorative plants that are its centerpiece. At the same time, a large portion of the recovered water is sent back to an underground reservoir that supplies the flush toilets with recycled, safe-to-use water. Applying this kind of thinking to the vertical-farm concept will allow for its application in nontraditional situations, including the reclaiming of drinking water from grey water, the manufacturing of pharmaceuticals extracted from higher plants, and the production of biofuels from algae and higher plants.

YOU CAN'T DRINK OIL

As a news feature reported in 2008, "The United Nations predicts that severe water shortages affecting at least 400 million people today will affect 4 billion people, more than half of humanity, by 2050" (www.pbs.org/newshour/extra/features/US/jan-june08/water_2-11.html). It is often stated that water is the new oil. I still have trouble with this one, since we need to drink 2.3 liters of water each and every day, but drinking oil? Well, I think what they really mean to say is that water is becoming more scarce by the day, and someday it may be even more valuable a resource than oil is today. It's also good to remind ourselves that more than 70 percent of the world's available freshwater is used to grow our food, an activity that renders it unfit for drinking. The world has to find alternatives to current water uses through the application of more efficient technologies, or we will suffer the consequences in the form of civil unrest and open warfare.

Cities are by far the biggest consumers of drinking water, turning it into black water (a combination of feces, urine, bathwater, runoff from storms, etc.) that each urban center then needs to handle and dispose of safely in order to prevent contaminating the local environment with its own waste. In the past, not being able to do so meant certain disaster for its citizens. Fecally contaminated drinking water has been responsible for outbreaks of cholera and dysentery, in which millions of lives have been lost. In the late nineteenth century, proof that many diseases were caused by microbes was closely followed by the invention of sanitation technologies that took advantage of this new knowledge, putting an end to horrible living conditions associated with an unclean environment in most European and North American cities. However, victory came at an enormous price, in terms of both dollars spent and human labor. New York City is a good example as to the extremes to which some urban communities have elected to go in order to ensure a safe and reliable source of drinking water. It will also serve as a case study here for what it might do to reverse its propensity for increased water consumption by recovering its grey water.

The Croton Reservoir and aqueduct system was designed to bring clean water to New York City to replace Collect Pond in lower Manhattan, a veritable cesspool of vile brown-colored liquid that was used by both people and draft animals. Started in 1837, the Croton system took five years to complete, revolutionizing city life in Old Gotham once and for all. One of the many benefits of an abundant freshwater supply was that the city need never again burn to the ground. Epidemics of cholera and dysentery also faded away shortly after its completion. As New York City's population burgeoned over the next fifty years, the city needed even more water, so a series of projects involving the construction of underground tunnels was conceived to bring water from much farther away, from the Catskill Mountains around 120 miles northwest of the city. Both branches of the Delaware River, the Schoharie River, the Esopus River, the Rondout Creek, and the Neversink River were dammed, and multiple tunnels were dug, a civil engineering feat that even the ancient Romans would have been proud of. Today, the reservoir system contains nearly 580 billion gallons. It had better, because 8.3 million New Yorkers consume an astounding 1 billion gallons of water per day. After using it for a variety of purposes, the city processes the combined sewage at fourteen treatment plants scattered around its five boroughs. The black water is then shipped by boat to Ward's Island, where it is dewatered by industrial-size centrifuges. The sludge part is processed further, removing the water and heating it to kill off all microbial life. It can then be powdered and used as fertilizer. The grey water created by the centrifugation process is returned to each treatment plant, again by boat, treated with chlorine, then unceremoniously dumped into the Hudson estuary. That's a lot of time and energy spent to move around boatloads of . . . well you know that rude word. Mostly, the biggest concern is the expense of the scheme. It generates little in the way of income (fertilizer aside), burns lots of fossil fuel, and wastes huge quantities of potentially valuable water. Scolding New York City for its wastrel habits seems a bit harsh, however, since nearly all cities, regardless of location, behave similarly. In fact, Washington, D.C., discards far more grey water than New York.

The Clean Water Act was created in 1972 under Richard Nixon's administration to put a stop to the dumping of untreated municipal sludge into the estuaries and oceans. The enforcement of its many regulations lies with the Environmental Protection Agency. The EPA requires that each community be solely responsible for discarding its wastes in ways that do no harm to the environment. As might be predicted, the implementation of modern waste-management practices has proven economically difficult for most cities to execute according to the letter of those laws, and equally expensive to sustain. Despite many helpful government-sponsored incentive programs that share the cost of converting older treatment plants into modern ones, U.S. cities collectively still spend billions of dollars each year disposing of liquid municipal waste. It's high time that we embrace the concept that these products of our own metabolism have intrinsic as well as economic value, and reverse that trend. To quote a tired old phrase, we need to turn lemons into lemonade.

For recovering the liquid part of black water, higher plants have the answer. As explained earlier, plants obtain their nutrients by pumping water through their roots, up through their leaves, and eventually transpiring it out into the atmosphere. Remediation of grey water could easily be accomplished by taking advantage of the enormous amount of transpiration that could occur inside vertical farms constructed expressly for that purpose. Dehumidification of the indoor air is all that would be necessary to get back the water we produced by eating and drinking. Can this idea actually work? Is it practical? How expensive is it? Would anyone in his right mind really drink water he knew came from a waste-treatment plant? As of this writing, there is no vertical farm devoted to water recovery, so predicting how much one might cost and how practical the idea actually is is hard to judge. However, social marketing of the general concept has already been accomplished: In Orange County, California, an initiative dubbed "from toilet to tap" is an example of recycling that emulates the best qualities of an eco-urban environment. For around $500 million, the city of Santa Ana constructed a processing plant that takes all the liquid in its municipal waste stream and recycles it. The people in that part of California, when asked, flatly refused to even consider drinking the water that came right out of the processing plant, even though they were well aware that there was no threat of a health hazard in doing so. The idea of drinking their own wastewater was, in their words, "Repulsive." (I'm glad the scientists working on the International Space Station didn't voice a similar objection or there would be no space station, but that's another story.) The citizens of Santa Ana instead chose to use that water to recharge the local aquifer, and then draw it out of the ground. The final filtration step was the earth itself. In contrast to that very elaborate—and, I might add, expensive—scenario, collecting transpired water from a building devoted exclusively to that process would be simpler, more direct, and most likely far less expensive. In addition, the plants would feed off the dissolved nutrients in grey water and grow. Harvesting excess plant material at periodic intervals would allow for the generation of energy by incineration, yielding an added bonus to the overall process. When it comes to whether this kind of vertical farm is practical (i.e., economically feasible), I would counter with, what expense would you go to in order to ensure a sustainable source of clean drinking water? Digging City Tunnel No. 3 under the state of New York's countryside to bring more water more efficiently to Gotham is a $6 billion plus solution that will be completed in 2020. I'd be willing to bet that making the vertical-farm water-recovery plan work would cost considerably less, it would be up and running light-years sooner, and, finally, it would be a local industry that would also help to clean the city's air by its very nature of being engineered by the plants.

INDOOR DRUGSTORE

The human species has been around for some two hundred thousand years, and over that time, due largely to geographic separation, we divided up into many cultures. Virtually all of them evolved a common set of survival tactics based on essential needs. Securing a reliable food supply was one of those, which this book has addressed from many perspectives throughout the last seven chapters. Dealing with illness was another aspect of life that all cultures had to grapple with, or else succumb to the forces of nature. Survival meant that they were either lucky, or that their immune systems were diverse enough to ward off the offending pathogens. Another possibility was that they had somehow stumbled upon something that aided in the healing process. When an individual became sick, attempts to alleviate suffering and identify the cause were natural outgrowths of our newly acquired ability to consciously reason through a given situation, and our propensity for altruistic behavior. Healers in those early settlements became the sages of their communities. In what must have been an arduous, gut-wrenching set of "preclinical" trials, using whatever they had at their disposal, humans gradually invented the science of therapeutics.

In the beginning, all healers had to rely on was the natural world. In many places this is still the case; witness the use of animal poultices (pieces of flesh or skin from either a fish, bird, amphibian, or mammal applied to an area of inflammation, such as an infected wound) and a plethora of herbal medicines, many of which were taken orally. What's more, in the majority of cases these natural products actually worked. In the case of animal poultices, scientists have isolated and studied the active ingredients. Most animal tissues contain a set of small peptide molecules related to the antibiotic gramicidin. These molecules can be routinely isolated from saliva, tears, urogenital secretions, frog skin mucus, and a variety of other sources. Quinine, a cure for malaria, and salicin, a pain reliever related to modern aspirin, are both plant-derived products still in use today. The short list includes digoxin, paclitaxel, reserpine, vinblastine, morphine, and many others. One can easily imagine how they all became incorporated into the shaman's repertoire of therapeutic agents. Both animals and plants of a given locality played important roles in defining the spectrum of active compounds derived from them. Because of the ready availability and abundance of plants compared to wild animals (and ease of capture, too, I might add), many cultures derived a remarkable number of useful drugs from herbs, shrubs, and woody plants that now comprise an extensive natural pharmacopoeia. Growing essential herbal plants from all the major cultures in a vertical farm devoted to that purpose would be most welcomed, especially in light of the fact that shortages of drugs encourage unscrupulous dealers to manufacture fake substitutes and flood the market with them. In addition, many of these plants are rare and may be harvested to near extinction in the name of humanitarian causes, leaving no options for treatment.

In 1828 Friedrich Wöhler synthesized the first organic molecule, urea, and set the world of industrial chemistry on its way. The modern science of dye chemistry, in turn, led the way to the establishment of the commercial drug industry around the nineteenth century. Aspirin is a synthetic form of acetylsalicylic acid. The Bayer Company in Germany synthesized it in 1897, and from that point on revolutionized the way medicine was practiced. The parent compound of acetylsalicylic acid is found in the leaf of the white willow tree (*Salix alba*), and has been known for its pain-relieving properties for centuries by native peoples wherever this species occurs. The pharmaceutical industry, founded on the same principles of drug discovery as those pioneered by Bayer, soon became a major economic force in the early twentieth century Industrial Revolution, and remains today as one of the major driving forces in the technosphere. Before there were standardized drugs available, all therapeutic agents had to be derived from natural sources, mostly from higher plants. According to the World Health Organization, of the 252 drugs considered as basic and essential, 11 percent are exclusively of plant origin, and a significant number are synthetic drugs obtained from natural precursors. All of the parent plants, save for those derived from trees, can be grown efficiently in a controlled environment. In addition, there are a few drugs that are derived from plants that the world routinely runs out of each year. One of them is artemisinin, which is the only effective treatment for those infected with drug-resistant malaria. Derived from the herb *Artemisia annua*, which grows wild in parts of Thailand and China, it is a drug in great demand, especially when world supplies get low. Dr. Facundo Fernandez, a professor of chemistry from Georgia Institute of Technology, conducted surveys of samples of artemisinin in 2009 by applying mass spectroscopy to confirm the contents of each vial of the substance. He found that the majority of them were devoid of any antiparasitic drug, and concluded that as a result of this chronic shortage, there is a brisk trade in fake artemisinin. Instead of the real drug, containers of bogus artemisinin were composed of a pain reliever like acetaminophen or ibuprofen, agents that treat only the fever part of malaria's symptoms. Of course, the unfortunate user does not know this and assumes he or she is getting better once the fever breaks and the pain is gone. Tragically, many victims of this scam die due to the lack of specific treatment. Illegal drug trafficking in artemisinin can be prevented by growing the plant in the vertical farm.

The Indian subcontinent is home to at least seven major religions and three systems of traditional medicine, one of which is the Ayurveda. The Ayurveda recommends 315 herbal medicines listed as essential drugs, with more than forty-two hundred currently registered plant derivatives. This ancient medical system was first created some five thousand years ago, and over that time has recommended the use of more than seventy-five hundred different plants for virtually anything that can afflict humankind, from infectious microbes to psychological problems. The term "ayurveda" translates roughly to "life sciences," and it continues to this day as the main reference for therapeutics from that vast region of the world. Cultures indigenous to Asia, South America, and Africa also have extensive lists of plant extracts for treatment of all human illnesses.

ENERGY IN, ENERGY OUT

Biofuels are best defined as combustible by-products of plant metabolism, such as oils used in biodiesel products, or microbial fermentation products of sugars derived from higher plants such as ethanol. Biodiesel can also be made from oils produced in lower plants like algae. In this case, specially constructed vertical farms would be needed. The production and use of biofuels have received much attention from big agri-business, with many brands of gasoline now boasting a 15 percent ethanol content at the pump. Brazil generates more than 30 percent of its automobile fuel by fermenting sugar pressed out of sugarcane, and is the world's largest producer of ethanol. Ethanol produced from corn can also be made by a chemical fermentation process. The upside to biofuels is that they are carbon neutral. Burning them generates carbon dioxide that is then captured by the plants and made into a potential biofuel product. On the negative side, the price of oil has forced farmers in many regions of the world to switch to corn and sugarcane for fuel production, lessening the amount of land devoted to food production. This has, in turn, resulted in higher food prices and food shortages in some places. Despite this, energy experts predict that clean-burning biofuels will eventually replace most of the fossil fuel–based petroleum products used in today's combustion engines. One solution to the land-use question is to grow all biofuel-producing plants in vertical farms.

THE PORTABLE VERTICAL FARM

Making food available for everyone is the big idea behind the concept of producing food in the vertical farm. In particular, people who are forced to migrate away from their place of origin due to war, civil unrest, or natural disaster would benefit most from having access to them. Clean water and a safe source of healthy food are the two things that the vertical farm can easily supply. In addition, displaced people need to settle somewhere in which good sanitary practices exist, or they may suffer even more insult than they did in the place they escaped from. Establishing a portable version of the vertical farm would apply to a number of these cases. Materials now exist—graphite composites for structural units, transparent lightweight window materials (e.g., ETFE), and the like—to make this initiative feasible today. Fast-growing, iron-rich leafy vegetables (spinach, kale) would be first-line crops, followed by a variety of root vegetables, protein-rich beans and grains, and the like, depending upon the ethnic preferences of the groups in question. Just as the MASH (mobile army surgical hospital) units, implemented first during the Korean War in 1951, revolutionized the practice of combat-zone medicine, portable vertical farms (PVFs) will restore the nutritional status of perhaps millions of disadvantaged, displaced people.

CHAPTER 9

FOOD FAST-FORWARDED

When you're finished changing, you're finished.

—BENJAMIN FRANKLIN

HERE TODAY, GONE TOMORROW

The concept of a disruptive technology is simple: It disrupts the present and jump-starts the future. The vertical farm has the potential to do that by advancing agriculture to a place in history it has never before occupied, one of true sustainability. But as with all new ways of doing things, there are some fairly hefty boulders in the middle of the road that need moving first. To begin with, bringing a vertical farm into reality, even a prototype, will require many elements to come together to permit its maximum expression. I am convinced that there already is a worldwide critical mass of political will, social acceptance, clever engineering, great designs, and science-based controlled-environment agriculture to coalesce the concept of vertical farming into a highly efficient food-producing building. When up and running, vertical farms will be yet another example of how the human species can solve problems that they themselves have created. Urban agriculture will lead the way to the establishment of a global network of functional food production systems situated directly in the mainstream of a crowded world, allowing for the repair of many of the world's damaged ecoystems. Despite all the enthusiasm for the concept, however, other problems will require immediate addressing if it is to succeed.

YOU CAN'T EAT MONEY

The most important unresolved issue comes from our current obsession with making lots of money as the only reason for introducing a new way of doing something. To always show a profit at the end of the day, no matter what we do, should not be the only motivation for innovation. There are some things that we all need for which making money is the furthest thing from our minds; for example, all of the public health services most countries provide free of charge. Well, not exactly free. Taxpayers pay for them. Lots of other essential services will also never make money, but we must have them in order to live better lives. Government support for farmers falls into that category. The pharmaceutical industry, a profit-driven enterprise if ever there was one, solves the profitability issue in a different way: diversification. All the successful drug houses occasionally discover a compound that is so useful that its sales alone make up for all the research-and-development costs for hundreds of other therapeutic agents that make little or no money at all. It's called the "cash cow" effect. When patents run out, companies panic and redouble their efforts to get another cash cow through the pipeline and out the door. Car companies, similarly driven by consumer preferences, make most of their revenue selling a mainline product so they can also offer high-end cars at luxury prices. All this might be changing in the wake of the economic breakdown of 2008.

So, now let's get serious about farming. Since when has farming ever been profitable? Okay, there are a few farmers who make millions growing hay in Alberta and Montana, and a few corn/soybean farmers in the American Midwest who do fine, as well. But throughout most of the world, farming is at best a break-even occupation. Want to see someone laugh until their sides hurt? Ask a small independent farmer what they earned last year. Want to see that farmer cry? Ask them the same question or, better yet, ask them to recount the reasons for the many crop failures they've endured over the last ten years. I think the real issue regarding the invention of vertical farming is, who will pay for the first ones? Answer that question and you will know who owns them. Part of the initial cost might involve the purchasing of a site. Real estate inside the city limits is usually not cheap. But, depending upon whom the project is intended for, the land might be donated, free of charge. Venture capitalists rarely give anything away, especially land, so as a group they are at the bottom of my list of potential funders. The same is true for banks or foundations with deep pockets. On the other hand, city, state, and federal governments are prone to encouraging the development of new areas of commerce by, for example, underwriting the building of a new sports arena or entertainment center, creating an industrial park by offering long-term favorable tax incentives, or donating city properties for mixed-income housing developments. These "sweetheart" deals make it nearly impossible to turn one down if the right partners are involved.

WE ARE FROM THE GOVERNMENT AND WE ARE HERE TO HELP. NO, REALLY.

In their most altruistic expression, governments exist to oversee the well-being of their citizens. Taxes are collected, then disbursed to create institutions for education, public health, social security, and the military and to ensure a reliable food supply, among other things. In a democracy, all of these governmental activities require buy-in from the majority of its stakeholders. Agriculture is the one area that every government, regardless of type, has never had a problem with convincing its citizenry to support. We all need to eat, and farmers are our indispensable lifeline to the supper table. In America, the public has never failed to financially back farmers' efforts at the congressional level, although the American farmer has managed to survive some very rough times without any help from Washington. Even our attitude toward the large agricultural conglomerates was rather benign until very recently, but times have changed. A recent disclosure by the Department of Justice revealed that less than 2 percent of farms account for half of all agricultural sales. Sounds like a monopoly to me. Remember also that the Food, Conservation, and Energy Act of 2008 was passed essentially unopposed to the tune of $280 billion. One percent of that appropriation could have gone a long way to developing the vertical farm from idea to reality (but that is wishful thinking on my part).

SO HERE'S MY PLAN

The USDA has amassed a remarkable farm-based infrastructure over the last seventy-five years, with regional offices in every state, successfully pushing for the permanent establishment of federal monies for crop insurance and supporting the establishment of land-grant colleges and universities for generating new technologies in all major areas of the food sciences. The main programs of the USDA over the last fifty years have succeeded beyond even their expectations, due mainly to the innovations associated with the second green revolution—pesticides, fertilizers, herbicides, and genetically engineered crops. Food availability is the USDA's proudest achievement, and most of it is for sale at reasonable prices throughout the lower forty-eight states. (Alaska and Hawaii have land-use and shipping issues that decrease consumer access to fresh produce and also increase the price of anything that has to be transported to those remote American outposts.) What if the U.S. federal government were to decide to create a whole new economic engine, in which urban agriculture was to become the centerpiece for a totally revamped model of how cities function, a model filled with lots of new job opportunities that fully embraced the concept of do no harm to the environment? How might it proceed if there were no political impediments or obsolete zoning regulations to prevent the emergence of a third green revolution?

I will now share with the reader a part of my consciousness normally reserved for fantasies involving catching the biggest trout of my life, or running the marathon in record time, or hitting the Powerball lottery for 300 million big ones. I invite you to also suspend your own sense of reality and imagine along with me that I am now in control of the country's purse strings. Without funding, the vertical-farm concept will simply disappear into the government warehouse featured in *Raiders of the Lost Ark* and get put up on a shelf alongside all the other good-intentioned ideas that failed due to lack of momentum and financial support. Here, then, is my master plan, complete with rationale for establishing a permanent presence for urban agriculture in every state in the union, with the vertical-farm concept as its focal point.

YOU CAN SOLVE A LOT OF PROBLEMS WITH FOOD AND WATER

Establishing many experimental vertical farms simultaneously is key, as it will enable a wide variety of like-minded scientists, architects, planners, engineers, and builders to enter into a collegial competition to push the vertical farm to its limits and beyond. How many groups should initially be established is a subject that could generate an entire day's worth of discussion among any governing body interested in getting the idea into the hands of those who can actually make it happen. In my fantasy, and please do keep in mind that it's simply a fantasy, I will empower every state to take the lead in developing its own version of the vertical farm. I believe that competition is healthy and productive as long as everyone competing has an equal chance of winning. My single objective as the person in charge of dishing out the dough will be to enable each group (soon to be defined) to design, build, and staff a prototype vertical farm. I don't want any of them to have to waste much time applying for grants and such. The money will be there, and identifying the participating members of each group will be up to a select committee established in each state. I have to be careful to structure my plan in such a way as not to discriminate against any state or major consumer group. I will accomplish this by supporting existing institutions and free up their time and enlist their creative energies with generous helpings of cash.

Each urban agricultural center (fifty in all) will receive a $100 million grant to be spent over a five-year period. Each state will agree to accept this federal money if they also agree to provide for personnel benefits, including health insurance and retirement packages in line with that state's standard employee benefits agreements. There will be stringent rules of engagement to ensure the proper conduct of business, complete with an oversight committee that will routinely audit the books for each center. I think the best model as to how to proceed is as follows: Each state will form a blue-ribbon committee for the purpose of choosing the actual members of the urban agricultural center that will manage the grant and do the work. The founder committee should be cochaired by the governor and the president of the most respected and productive university in that state that also is engaged in ongoing agricultural research. Other members include the deans of the schools of agriculture, architecture, engineering, biological sciences, business, and public health (wherever possible). Once they have selected the working group, their job is done. The next steps would be up to the new committee to establish ground rules for proceeding to design, build, and manage the prototype vertical farm. Under the rules of the grant, all building should be completed by year two. Each group is free to select the specific crops to be grown inside, but they must include a balance of types of edible plants—fruits, vegetables, grains, herbs, and spices. The managing of each enterprise would be at the university level, perhaps in a special division in the university's office of grants and contracts. The prototype could employ a total of several hundred people. The corporate structure will be left up to each group. Some may choose to go the academic route, establishing a department of urban agriculture, while others might opt for something more elegant in name; perhaps calling themselves the Center for Urban Sustainable Agriculture.

Other government-funding strategies already exist and

could easily be applied to facilitate the creation of the vertical farm and other related versions of urban agriculture. A request for proposals to establish a prototype vertical farm would be issued for the purpose of funding ten centers of excellence. Not every state will enter the bidding, but all of them will eventually benefit from the execution of the five-year grants by the centers that do get awarded the $100 million. This model has in the past proven highly effective in encouraging the development of a good number of high-profile federally sponsored research programs, including cancer treatment and prevention centers, high-energy particle physics facilities, telescopes, and biomedical-research centers focusing on neurological diseases. If this program catches fire, then I would encourage another round to make a total of twenty vertical farms, spread evenly throughout the country.

In any case, when the first prototypical vertical farms are up and running, I would then establish a number of annual public- and private-sponsored competitions with big cash prizes for things like best overall design; best energy-saving system; best use of passive energy; best system for recycling water; best use of automation for monitoring pH, root temperature, and nutrient delivery in a hydroponic system; and best yield per square foot for wheat, chickpeas, barley, rice, and other essential crops. Other prizes would be given for the best-tasting tomatoes, most innovative use for hydroponic and aeroponic growth systems, and the like. The winning teams would share equally in the distribution of monies, rewarding everyone who participated in the vertical farm (as in the World Series, where even the bat boy of the championship team gets a share). The anticipated result of all this activity would be the permanent establishment of urban agricultural initiatives throughout the entire country. Each group would be obliged to partner with a city in its state willing to be the urban laboratory for trying out practical applications of its research findings. Restaurants, schools, hospitals, apartment complexes, and senior-citizen facilities would all be fair game for attaching a version of the vertical farm to them. Free-standing vertical-farm complexes, complete with all the ancillary buildings described in chapter 6, would be the culmination of years' worth of applied research defining the limits of this new agricultural strategy. They would rapidly become showplaces for sustainability. A significant spin-off from the program would be the establishment of a national Society for Urban Sustainable Agriculture, a professional research journal, and an annual meeting, at which graduate students and their mentors could exchange findings in a series of ten-minute talks, poster sessions, and plenary lectures. Monographs and books on the subject would undoubtedly follow, as academic careers matured and the science behind the practice of vertical farming became more and more focused.

Having spent many years in academia, I am confident that the research part of the plan I have roughly outlined would have a good chance of succeeding, based on a series of other highly successful initiatives that I am familiar with in the biological, physical, and chemical sciences. Research carried out by universities and colleges is now a main conduit for the commercialization of scads of new technologies; witness Silicon Valley, et cetera, et cetera, et cetera (apologies to Yul Brynner). Despite an enormously successful technology-driven college-based research enterprise, some business types might argue that my plan, as sketchy and optimistic as presented, is rife with flaws of all kinds, mostly financial ones. I tend to agree with them, but we have to start the discussion somewhere in scoping out how to proceed. Other models for getting the vertical farm literally off the ground would require huge inputs of money from the private sector, but as I pointed out earlier, profit is the key here. The vertical farm will undoubtedly make money and lots of it, but the return on investment may take a bit longer than most investors are willing to wait. Failures in design and execution have to be paid for, and no one wants to end up owning a drawerful of reports documenting mistake after mistake with little else to show for his or her time and effort.

TO DREAM THE IMPOSSIBLE DREAM

Well, come to think of it, maybe some businesses do take risks. Despite recent setbacks, the Toyota Motor Corporation is a car manufacturer with vision and lots of resources. It has the luxury of success on its side, and therefore it can afford to take chances that might have big payoffs. Toyota took a very big one in developing a new kind of automobile, reasoning correctly that consumers would buy a car that burned less fossil fuel and got amazing mileage. Enter the hybrid vehicle. Only one thing stood in the company's way: a reliable, cheap, small, high-energy rechargeable battery. Not a new idea, but when it announced its plan to the world the idea was within reach from a research perspective. Someone had to be first and invest in the technology. Toyota had bravely stepped forward and seized the lead. The rest of the international automotive industry shook their heads and laughed at the idea, continuing to manufacture gas guzzlers, and completely misjudged or downright ignored the consumer demand for a greener ride. The Bush administration was no help at all, and in 2001 discontinued federally sponsored research and development of hybrid vehicles, a program first initiated by President Clinton. In 1995 Toyota announced that it had finally succeeded, and the first Prius stealthily purred off the assembly line and into the international auto shows as a concept car for all to see. In 1997 Japan began selling them in its own country, and by 2000, models were being sold in the United States. The U.S. buying public ate them up, driving them out of the showroom and into the hearts of the world's environmentally conscious drivers. In the Prius's first year of full production, 2003, Toyota could not keep up with demand, and as everyone now knows, the rest was history. A new version now even comes with a solar panel on the roof to offset even more fossil-fuel use. A long line of auto manufacturers formed at their doorstep to get on board the new gravy train. The behind-the-back snickering and joking turned to apologetic bows of respect at the signing ceremonies for licensing the battery technology. Toyota ended up with a pocketful of gold. The other automakers ended up saving a bundle by not having to invent something they could now buy. Everybody won, including the consumer and the environment.

Are there similar forward-looking companies in the United States that could take up the challenge of helping to invent the vertical farm? In 2009 *Fortune* magazine listed among its top ten (NetApp, Edward Jones, Boston Consulting Group, Google, Wegmans, Cisco, Genentech, Methodist Hospital, Goldman Sachs, and Nugget Market) several that might fall into this category. Google would be my first choice. This giant has an altruistic streak a mile wide. Google could afford to promote the concept with significant financial aid. So, who will be first to invent the vertical farm? I can hardly wait to see!

AIM HIGH

I am going to assume that within the next few months, a serious initiative to build a vertical farm will be under way. I base this prediction on the overwhelmingly enthusiastic response the idea gets every time I speak, and on communications to me and my colleagues via the Internet. So, given that there will eventually be vertical farms, what are the long-term implications of urban agriculture in the context of the future city? So far, I have purposely presented the concept in its best light without too much speculation and detail regarding some of the difficulties that might be encountered during its execution, save for funding. Negative thinking is not in my nature. I am an optimist through and through, and I remain thoroughly convinced that we can solve any problem, no matter how big or small. We simply have to want to solve it. Of course there will always be a handful of detractors voicing their negative-laden critiques for any idea, no matter if it's an important one or not. I like to keep in mind the lyrics of George Gershwin's American Songbook classic "They All Laughed." So go ahead and have a good chuckle if you must, but time is running out. The prototype vertical farm is right around the corner and it will become the proving grounds for its eventual success.

The real question is, what lies ahead once all forms of urban agriculture take hold and the city begins to feel the positive effects of in-house food production? For a clear vision of urban life over the next fifty years, let's go back to a more complete discussion of what the eco-city might look like. Bioproductivity is key to its success. Vertical farms and rooftop greenhouses have now solved that problem. So, with food-generating systems in place, the metropolis can finally turn its attention to the real task of creating a varied and interesting landscape, providing its inhabitants with healthy alternatives for living regardless of location and the specter of social and environmental injustices related to a deteriorated neighborhood. The rich and the poor need to find virtue in the landscape they live in in order for the city to fulfill its manifest destiny as a place where we all want to live and contribute to its institutions.

AIN'T MISBEHAVIN'

The eco-city of tomorrow will have a built-in code of social ethics based on principals dictated by natural law. The most important one will be that a city cannot exceed its energy limits for any reason. Encroachment into natural systems to secure more resources will not be tolerated, since we will have once and for all learned our lesson regarding unwanted negative consequences for going where we are not welcome. Hence, the rational management of energy will become the number-one priority. Passive energy capture will be commonplace, and regulations encouraging all buildings to strive toward a zero-carbon footprint will be the mantra of city councils, the mayor's office, and other governmental systems around the country. Elections will be won or lost depending upon how well those in power adhere to this ecological golden rule. It is possible to achieve this elusive goal with the massive application of known technologies. Some shining examples include things already mentioned; plasma arc gasification, for one. Along with eliminating the need to deal with municipal solid and liquid waste disposal, the city needs to become more aware of public transportation issues. Electric cars, safe bicycle paths, and a host of other people-friendly programs to encourage the use of all public spaces will unite its residents into a coherent body of voters whose sole aim will be the continued improvement of their environment. It is really not such a big stretch to imagine what it might be like to live in the eco-city of tomorrow from where we find ourselves today. There are a host of new programs scattered throughout the world that, when taken as a whole, could very well define how all of the successful green initiatives might come together to reform life in big cites.

WE HAVE THE WHOLE WORLD IN OUR HANDS

In the end, the choice is ours as to whether or not we will carry out our lives in an ecologically responsible fashion. Throughout our history, the human species has adapted to a changing environment by inventing technological solutions that made our life easier: better housing, more productive farming methods, global transportation, on-demand communications systems, and a host of medical intervention strategies. All of these activities have had both positive and negative effects on the way our planet functions. The time has arrived for humans to reassess their place in the natural world, embrace and celebrate the differences between us and the rest of the creatures that comprise it, and incorporate a reverence for our origins that reflects itself in a new respect for the DNA molecule, no matter what form it takes. In doing so, we will have achieved a major milestone in our evolution: sustainability into the millennium level.

Courtesy of *diephotodesigner.de*

An Infarm grower nursing plants in farm trays

Growers working at an Infarm InHub

Courtesy of *diephotodesigner.de*

Courtesy of diephotodesigner.de

Growers working at an Infarm InHub

Courtesy of diephotodesigner.de

A close-up of a farm tray at an Infarm InHub

Courtesy of Spencer Lowell

Courtesy of Spencer Lowell

(above) A Plenty farm operations associate walking past the transplanter

(left) A Plenty farm operations associate in front of leafy greens growing in vertical towers

Courtesy of Spencer Lowell

A wide shot of the central processing area at Plenty

A view of the Plenty grow space through the vestibule window

Courtesy of Spencer Lowell

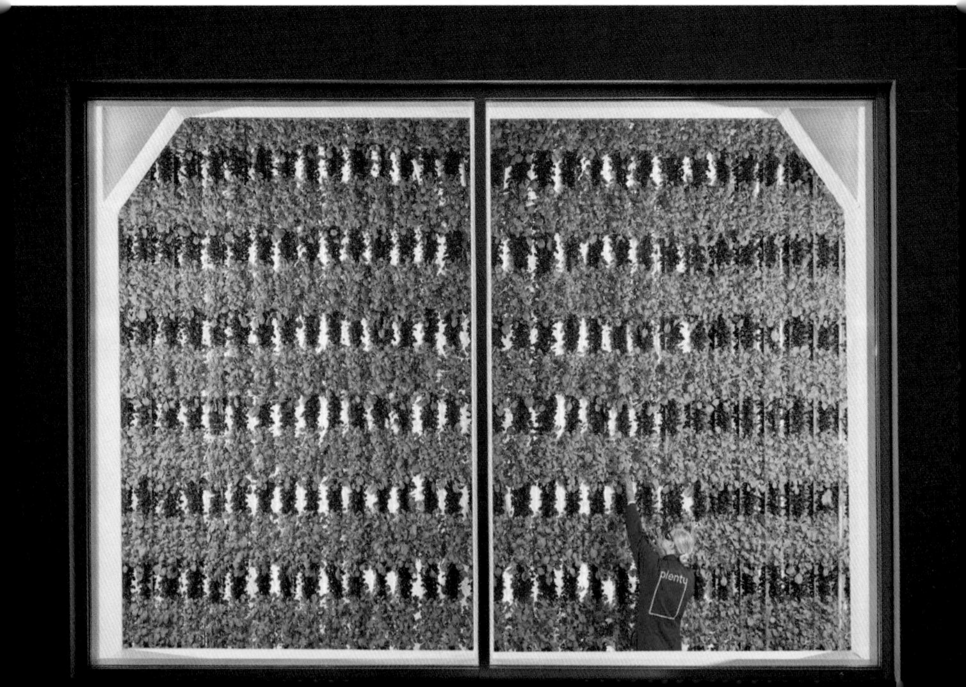

Courtesy of Spencer Lowell

Plenty farm operations associates in the grow space

One of Bowery Farming's growing systems in Kearny, New Jersey

Courtesy of Bowery Farming

Courtesy of Bowery Farming

A close-up of plants growing at Bowery Farming's Kearny farm

Plants growing at Bowery Farming's Nottingham, Maryland, farm

Courtesy of Bowery Farming

Courtesy of Vertical Harvest

Courtesy of Kalera, Inc.

(above) The exterior of the Vertical Harvest farm in Jackson Hole, Wyoming

(left) The exterior of Kalera, Inc.'s HyCube facility in Orlando, Florida

(below) The exterior of the AeroFarms facility

Courtesy of AeroFarms

Courtesy of AeroFarms

A worker tending plants in the AeroFarms facility

CHAPTER 10

THEN WHAT HAPPENED?

In 2010, when this book was first published, there were no vertical farms as far as I could determine. Three years later, there were three: one in Suwon, South Korea; one in Japan; and one in Bedford Park, Illinois. Since then, the idea has blossomed, most likely because recent progress on many industrial fronts has lowered the startup costs and thus made owning and operating profitable. As of this writing, there are so many vertical farms that I don't know exactly how many exist. Some of the largest vertical farms include the Jones Food Company in England; AeroFarms in Newark, New Jersey; Plenty in San Francisco; Oasis Biotech in Las Vegas; Green Sense Farms in Portage, Indiana; Emirates Airline's farm in Dubai; and Spread in Japan. For a select list of farms, see Table 1.

TABLE 1.

SELECT VERTICAL FARMS

FARM	LOCATION	WEB SITE
	United States	
AeroFarms	Newark, NJ Danville, VA	aerofarms.com
Bowery Farming	Kearny, NJ Nottingham, MD	boweryfarming.com
Cool Greens	Alden, NY	N/A
FreshBox Farms	Millis, MA	freshboxfarms.com
Ceres Greens	Barre, VT	ceresgreens.com
Metrocrop	Bridgeport, CT	metrocrops.com
80 Acres Farms	Hamilton, OH	80acresfarms.com
Buckeye Fresh	Medina, OH	buckeyefresh.com
Balance Farms	Toledo, OH	balancefarms.com
Green Sense Farms	Portage, IN Henderson, NV	greensensefarms.com
Green Spirit Farms	New Buffalo, MI	greenspiritfarms.com
Up Farms	Chicago, IL	N/A
Ernessifarms	Ripon, WI	ernessifarms.com
Living Greens Farm	Faribault, MN	livinggreensfarm.com
Garden Fresh Farms	Maplewood, MN	gardenfreshfarms.com
Lettuce Abound Farms	New London, MN	lettuceabound.com
Infinite Harvest	Lakewood, CO	infinite-harvest.com

FARM	LOCATION	WEB SITE
Vertical Harvest	Jackson, WY	verticalharvestfarms .com
Oasis Biotech	Las Vegas, NV	oasisbiotech.com
Plenty	San Francisco, CA Kent, WA	plenty.com
Infarm (partnered with Kroger)	Seattle, WA	infarm.com
VolcanoVeggies	Bend, OR	volcanoveggies.com
True Garden	Mesa, AZ	truegarden.com
Urban Rebel Farms	Santa Fe, NM	urbanrebelfarms.com
Sananbio	Albuquerque, NM	sananbious.com /vertical-farming
Fifth Season	Braddock, PA	fifthseasonfresh.com/
Lancaster Vertical Farm Initiative	Lancaster, PA	N/A
Moonflower Farms	Houston, TX	moonflowerfarms .com
Eden Green Technology	Cleburne, TX	edengreen.com
MoFlo Aeroponics	LaGrange, GA	mofloaeroponics.com
Second Chances Farms	Wilmington, DE	secondchancesfarms .com
Kalera	Orlando, FL	kalera.com
	Canada	
Ahluwalia Farms	Edmonton, AB	freshfarms.ca
We the Roots	Toronto, ON	intravisiongroup.com
Living Earth Farm	North York, ON	livingearthfarm.ca

FARM	LOCATION	WEB SITE
	United Kingdom	
	Scotland	
Intelligent Growth Solutions	Dundee	intelligentgrowth solutions.com
	England	
GrowUp Urban Farms	London	growup.org.uk
Jones Food	North Lincolnshire	jonesfoodcompany .co.uk
LettUs Grow	Bristol	lettusgrow.com
Shockingly Fresh	Offenham	shockinglyfresh.co.uk
Vertical Future	London	verticalfuture.co.uk
Infarm	London	infarm.com
	Europe	
	The Netherlands	
Philips GrowWise Center	Eindhoven	lighting.philips.com
Staay Food Group	Dronten	N/A
	Germany	
Infarm	Berlin	infarm.com
	Denmark	
Infarm	Copenhagen	infarm.com
	France	
Florentaise	Saint-Mars-du-Désert	florentaise.com
Ilimelgo Architects	Paris	ilimelgo.com
	Belgium	
Urban Crops	Waregem	urbancropsolutions .com

FARM	LOCATION	WEB SITE
	Italy	
Planet Farms	Milan	planetfarms.ag/
	Norway	
BySpire	Oslo	byspire.no
	Finland	
Evergreen Farm OY	Tampere	evergreenfarm.eu
	Russia	
Mestnye Korny	Moscow	fruitnews.ru
iFarm Project	Novosibirsk	ifarmproject.ru
	Moscow	
	United Arab Emirates	
	Dubai	
Badia Farms	Dubai	badiafarms.com
APEX Insight	Dubai	apex-insight.com
Plenty	Dubai	plenty.ag
	Abu Dhabi	
Plenty	Abu Dhabi	plenty.ag
	India	
Urban Kisaan	Hyderabad	urbankisaan.com
Herbivore Farms	Mumbai	N/A
Terra Farms	Manori	N/A
	Singapore	
Sky Greens	Singapore	skygreens.com
Upgrown Farming Co.	Singapore	upgrownfarming.com
Vertical Farm	Singapore	N/A
Panasonic VF	Singapore	panasonic.com

FARM	LOCATION	WEB SITE
	Taiwan (100+ VFs)	
Yes Health iFarm	Luzhu	yeshealth.com
	South Korea	
NextOn	Dangjae Tunnel, Seoul (in subway station)	inexton.com
	China	
Farm66	Hong Kong	farm66.com/en/home/
	Japan (200+ VFs)	
Spread	Kyoto	spread.co.jp
	Australia	
Stacked Farm	Burleigh Heads	stackedfarm.com.au
Vertical Farm Systems	North Arm	verticalfarms.com.au

When the idea of vertical farms first arose in my class, we decided early on that in order to maximize food production to feed as many people as possible from a single building, the growing space would have to occupy at least several stories, in contrast to rooftop gardens and greenhouses. This is an important point to keep in mind, especially now that the vertical-farm concept has become widely accepted. Many self-proclaimed vertical farms were not in keeping with our original definition, since they were only single-story structures, and they are not included in this update. Nonetheless,

I fully acknowledge the many significant contributions to urban agriculture of rooftop gardens and greenhouses, and large-scale, high-tech rooftop greenhouse companies, such as Gotham Greens in Brooklyn and Chicago, and Lufa Farms in Montreal. All these efforts continue to increase public awareness regarding the feasibility and desirability of even larger scale agricultural iterations, including vertical farms.

Indoor agriculture attracted even more attention after 2015, and several Web sites went online featuring news and updates related to vertical farming and urban agriculture: the Vertical Farm (www.verticalfarm.com), Agritecture (www.agritecture.com), the Urban Vertical Farming Project (www.verticalfarmingproject.com), and iGrow (iGrow .news). Associations formed, tailored to the needs of vertical farm owner/operators, while many interested nonprofessionals also joined them. These organizations continue to serve as conduits of information regarding all aspects of the indoor growing industry: the Association for Vertical Farming (vertical-farming.net), the Vertical Farm Institute (verticalfarminstitute.org), the FarmTech Society (farmtechsociety.org), and the newly formed American Association for Urban and Vertical Farming.

A number of international trade shows started featuring manufacturers of entire indoor growing systems (hydroponic, aeroponic, and aquaponic); remote sensing instrumentation and computer-based technologies for monitoring plant growth and health; adjustable, adaptive scaffolding and related supplies (e.g., tubing, nutrient reservoirs, pumps) for creating vertical farms and high-tech greenhouses; a wide variety of high-efficiency LED grow lights; and plant-nutrient solutions tailor-made for a wide variety of crops. Many of the trade shows on indoor agriculture have become annual events. Among the more popular ones are Seeds&Chips in Milan, GreenTech in Amsterdam, AgriTech Summit in Utrecht, AgriFood Innovation Event in Venlo, Novel Farm in Pordenone, GFIA in Abu Dhabi, Indoor Agriculture Energy Solutions in San Diego, and Ag Innovation Showcase in Minneapolis.

Two airlines have partnered with vertical-farm companies to provide healthier in-flight meals for travelers. In Dubai, Emirates Airline formed an alliance with Crop One Holdings, and in 2018, commissioned the construction of a $40 million controlled-environment agriculture (CEA) facility that grows a wide variety of salad ingredients for its entire fleet of passenger planes. The following year, AeroFarms partnered with Singapore Airlines. It is anticipated that others will follow their lead within the next few years, providing more and more passengers with healthier, fresher meals.

The rapid growth of the vertical-farm (VF) industry captured the attention of a number of commercial marketing firms who now publish annual reports projecting the compound annual growth rate (CAGR) of the VF industry. The majority of those reports estimate that sometime within the next five to ten years, the global VF industry will be worth an estimated $13 billion. Virtually every one of these documents is sequestered behind a hefty paywall, preventing the average investor from gaining detailed financial analysis and insight regarding the pros and cons of owning and operating a vertical farm. This obstacle has not detoured the "boom economy bus" that is driving the establishment of vertical farms throughout the world. In my opinion, its remarkable early success is in large part attributable to the fact that given the advances in greenhouse-based technologies in over just the last three years, anyone who wishes to enter the field need not worry about technical barriers. The advent of high-efficiency, lower-cost LED grow light systems, flexible scaffolding for a wide variety of growing strategies, remote sensing programs to monitor plant growth and health, and a number of consulting firms specializing in advising vertical-farm startups have provided enough incentives and encouragement to overcome any concerns as to how to come up to speed. The great majority of owner/operators of vertical farms are ambitious, enthusiastic, and knowledgeable millennials; many are under the age of forty. To them, vertical farming is their way of remaining within their communities, focusing on producing food for their fellow city dwellers.

Some vertical farms have embedded social programs into their corporate mission statements. For example, AeroFarms, in collaboration with the mayor's office of Newark, hires and trains local unemployed young men and women, helping to alleviate chronic unemployment that has plagued that city for the past few decades. Vertical Harvest, located in downtown Jackson, Wyoming, with partial financing from a grant from the state, successfully implemented a plan to employ and train individuals with a variety of learning disabilities, including autism and Down's syndrome.

At present, establishment of new vertical farms seems limited only by the number of abandoned warehouses and obsolete manufacturing facilities. There is an abundance of these structures throughout the built environment, regardless of global location, making the retrofitting of these buildings the primary strategy for a rapid portal of entry into the VF industry. Construction of a new multistory growing space presents a more daunting financial model. In either case, fundraising is straightforward, provided that the newly formed company can provide potential investors with proof of concept. Prior to seeking major funding streams, pilot studies are an absolute necessity. This allows time to work out the detailed protocols for planting, growing, harvesting, packaging, and marketing. Once the routine for growing the desired crops has been established, funders can be invited in to see the fruits of their labor, quite literally. Nothing makes a stronger case for convincing funders that you will succeed than having them eat a tasty salad of mixed greens and vine-ripened tomatoes harvested less than an hour earlier. Using this approach, AeroFarms, headed by David Rosenberg, Marc Oshima, and Ed Harwood, garnered an initial investment of $80 million in funding from Goldman Sachs, the city of Newark, and Prudential Financial. Plenty, a San Francisco–based vertical-farm company (Matt Barnard, CEO), secured an even larger investment of $226 million from SoftBank of Japan and from Jeff Bezos, CEO of Amazon. Similar success stories abound in the VF industry and speak volumes about the enthusiasm for this kind of technology among certain investor groups, as they seek new ways of supporting the creation of a more sustainable, livable urban environment.

In addition to business acumen in crafting long- and short-term financial plans, marketing, sales, human resources, and resource allocation of initial investment funds, an intimate working knowledge of which crops will be grown and in what kind of growing mode (e.g., hydroponic, aeroponic) they will be produced is central to the technology side of vertical farming. As of 2020, there was an acute shortage of qualified indoor agronomists. This is probably the most important limiting factor affecting the rate of growth of the industry. There is every reason to believe that this situation will soon change. Governments, city planners, and academic institutions are becoming more aware of this new industry and are beginning to create training programs for those who want to become city farmers.

The United States Department of Agriculture has established the Office of Urban Agriculture and Innovative Production, funded with $5 million added to the 2018 Farm Bill. In addition, it also established the Urban, Indoor, and Other Emerging Agricultural Production Research, Education, and Extension Initiative, which will receive $4 million per year through 2023. This initiative authorizes competitive grants to support research, education, and extension activities for the purpose of facilitating the spread of urban, indoor, and other emerging agricultural production. The bill also requires the Secretary of Agriculture to conduct a census of urban, indoor, and other emerging agricultural production.

The cities of Shanghai and Rotterdam have initiated the establishment of CEA training centers to attract those who want to become part of the third green revolution. Universities offering complete educational programs in CEA include the University of Arizona, Cornell University, the University of Nottingham, and Wageningen University & Research. The University of California, Davis, has announced plans to construct state-of-the-art greenhouses to allow advanced research in CEA programs. Many other schools of agriculture are in the planning stage to establish programs in CEA. A large number of schools of architecture have regularly scheduled studios that emphasize the inclusion of some form of urban agriculture into the built environment. It is expected that due to the increasing popularity of this subject among precollege- and college-age students, within the next five to ten years, many more schools of agriculture, engineering, and architecture throughout the world will follow suit and become centers for the advancement of vertical and urban farming. Since activity in all of these areas is ongoing, the future of vertical farming looks very bright indeed.

THE NEXT THIRTY YEARS

Vertical farming will soon play a significant role in the evolution of the modern city by providing most of the food for those who live there. Vertical farming will greatly reduce the need to grow crops outdoors. As farmland is abandoned, it will return to its original ecological function by a process called rewilding. Carbon sequestration will increase as rewilding accelerates, and the rate of climate change will slow down.

Today, vertical farms' main commercial crops are mostly leafy greens. This menu is destined to expand into a full-fledged selection of edible plants, including grains, for two main reasons. First, consumer demand will inevitably convince the VF industry to accommodate consumers' desire for a one-stop-shop daily visit to the local vertical-farm supermarket. Second, rapid climate change will most likely continue to increase in severity over the next decade no matter what we do now, adversely affecting what crops can be grown, their yields, and where outdoor farming can occur. Food shortages will become the new normal as our population increases and coastal regions of the world decrease in area due to sea level rise. Hyper-urbanization will be the result, and we must address this impending crisis today if we are to avoid civil unrest and wars fought over food.

Within the next twenty to thirty years, nearly 70 percent of the human population will live in some form of urban environment. Even though cities have been in existence for some eleven thousand years—Damascus is credited with being the first—we still have not yet mastered how to live in them in a sustainable, resilient, equitable way, ensuring that every urban dweller is well-fed, has access to clean water and nonpolluted air, and can live in adequate housing. Homelessness is one of the strongest indicators I can think of to evaluate how livable a city is. Few can boast that they have solved this problem. Until all of them do, I will hold on to my worldview.

The modern city has an insatiable appetite for nonrenewable resources, and massive environmental degradation is the unintended consequence of that behavior. But there is another way of conducting business in which urban centers do no harm to the surrounding landscape. Two new technologies have recently emerged that, when applied to the design and construction of new buildings, and in combination with vertical farming and rainwater harvesting, will allow us to erect cities that respect and value nature instead of destroying it. We can create self-sufficient urban environments that essentially leave the natural world alone.

CARBON SEQUESTRATION

The global carbon cycle is out of control. According to the National Oceanographic and Atmospheric Administration, as of 2020, the level of atmospheric carbon dioxide has risen to 414 parts per million. This situation has occurred within just the last two hundred years, and is coincident with the advent of the industrial revolution. Meanwhile, the steady increase in the human population has resulted in the conversion of a huge portion of the terrestrial landscape into farmland—about 1.84 billion hectares of mostly hardwood forests and grasslands. Tree numbers have plummeted, going from an estimated 6 trillion prior to the agricultural revolution to 3.04 trillion.* Rapid climate change is directly linked to this finding, as trees play a major role in sequestering carbon, storing it in their trunks, branches, and roots in the form of cellulose.†

* T. W. Crowther et al., "Mapping tree density at a global scale," *Nature* 525 (September 2015): 201–5

† G. Popkin, "How much can trees fight climate change?" *Nature* 565 (January 2019): 280–82

To make matters worse, cities contribute significantly to the greenhouse-gas dilemma in several ways. First, the majority of cities regulate the construction industries, requiring them to use concrete, steel, and other inert materials that have incredibly high energy footprints associated with their manufacture. The mining and smelting of ores are carbon pollution–intensive activities. Second, most cities use fossil fuels as an energy source to heat and cool their buildings, and in doing so are responsible for nearly 20 percent of the atmospheric carbon dioxide.* The good news is that all this could change if we want it to.

Cross-laminated timber (CLT), an engineered wood product is revolutionizing how we design and build.† The advantages of using engineered wood as a building material are many. CLT is a renewable resource, is twice as light and stronger than an equal volume of steel, can be easily disassembled and reused if necessary (concrete cannot be reused), and an entire building can be manufactured using computer-programmed laser-guided saws off-site, then rapidly assembled onsite within a matter of weeks, à la LEGO. In contrast, construction of modern skyscrapers takes months, and some take years to complete. Since wood is the

* D. Moran et al., "Carbon footprints of 13,000 cities," *Environmental Research Letters* 13, no. 6 (June 2018)

† https://www.archdaily.com/893442/cross-laminated-timber-clt-what-it-is-and-how-to-use-it; https://www.archdaily.com/922980/is-cross-laminated-timber-clt-the-concrete-of-the-future; https://www.ted.com/talks/michael_green_why_we_should_build_wooden_skyscrapers

source of CLT, trees that were cut down for CLT assembly will eventually be replaced by new growth, sequestering more carbon from the atmosphere.

The resiliency and safety of CLT buildings has come under the scrutiny of the building trades. Numerous studies have shown that CLT is difficult to burn.* Due to its thickness, the outer surfaces char but do not support long-term burns.

To date, hundreds of tall buildings have been fabricated out of CLT.† If engineered wood becomes the major building material with which cities replace old structures, cities have the potential to become a significant carbon sink over the next one hundred to two hundred years.‡

CLEAR PHOTOVOLTAICS

Glass is the accepted standard for modern skyscraper facades. By converting the surface of windows with a photon collection device (another emerging technology), a method referred to as semitransparent building-integrated photovoltaics (BIPV), all windows can become producers of elec-

* https://www.nfpa.org/News-and-Research/Data-research-and-tools/Building-and-Life-Safety/Fire-Safety-Challenges-of-Tall-Wood-Buildings-Phase-2

† https://urbannext.net/wooden-skyscraper/

‡ F. Lowenstein, B. Donahue, and D. Foster, "Let's Fill Our Cities With Taller, Wooden Buildings," *The New York Times*, October 3, 2019

tricity.* To date, the record for conversion of sunlight to electrons in windows coated with a wide variety of photon-capturing substances is low, about 15 percent (according to laboratory results reported from Michigan State University in 2019). But as with all new technologies deemed worthy of further research, efficiency will undoubtedly increase, and in a very short period, if history is any indication as to the time scale needed for achieving practicality.

There are several ways of converting a plate of glass to photocell mode.† Each method of coating glass with a photon-capturing substance has its advantages and disadvantages. One of the main issues so far centers on how much reduction in transparency is necessary in order to achieve practical levels of energy generation. Office windows should be at least 80 percent transparent for maximum work efficiency. Which one of the current technologies will succeed over the others, or will perhaps even be replaced by an entirely new approach to improving the efficiency of clear glass-based energy generation, cannot be predicted based on today's research findings. Several comprehensive reviews of these technologies are supportive of the expectation that clear photovoltaic windows will become an inte-

* R. Service, "Skyscrapers could soon generate their own power, thanks to see-through solar cells," *Science*, June 28, 2018

† P. K. Nayak et al., "Photovoltaic solar cell technologies: analysing the state of the art," *Nature Reviews Materials* 4 (March 2019): 269–85

gral part of the smart city in the near future. In their review of transparent solar photovoltaic technologies, Alaa A.F. Husain and colleagues concluded: "It is expected that this research will result in the integration of TPV in most electrical applications. Soon, mobile devices will self-charge, and skyscrapers will have zero net energy consumption without needing additional roof space for solar panels."*

A year later in another review of the same technologies, Benedicto Joseph and colleagues drew similar conclusions: "The semitransparent BIPV was demonstrated to have many advantages as a replacement of the conventional materials of building facades, while at the same time providing clean energy, thus reducing unclean external source dependency."†

A few successful commercial applications of clear photovoltaics have been achieved (e.g., Onyx Solar Co. and Ubiquitous Energy Co.). It is anticipated that the efficiency for commercially produced versions of BIPV (currently at 9.8 percent), cost, and robustness of the technology will soon make it a practical solution for generating most, if not all, of the energy needs of a city. It is even possible that cities may become the supplier of electricity for their surrounding suburbs, as well. The development of a long-term storage

* A.A.F. Husain et al., "A review of transparent solar photovoltaic technologies," *Renewable and Sustainable Energy Reviews*, 94 (October 2018): 790–91

† B. Joseph et al., "Semitransparent Building-Integrated Photovoltaic: A Review on Energy Performance, Challenges, and Future Potential," *International Journal of Photoenergy* 2019 (October 2019): 1–17

battery will undoubtedly add a needed missing ingredient to propel the modern city toward energy grid independence that I have termed "Grid Lack."

RAINWATER HARVESTING

Bermuda is famous for its lovely, pastel houses that feature spotlessly clean white roofs designed to capture all the rainwater that falls on them. That water is used for everything from drinking to washing dishes. Bermuda has no groundwater reserves, so it must behave in this way if it is to survive, and survive it has. India is equally known for its penchant to take advantage of this "gift from heaven." Many cities on that subcontinent have ordinances that require all of their inhabitants to collect as much rainwater as possible that falls on their domiciles. This change in behavior over the last twenty years has been triggered by a shift in the monsoons that now dump too much water at the onset, then peter out, leaving little in the way of groundwater at the end of the yearly cycle for farming and everyday household needs.

These two examples are proof enough that it is not only possible for any city to capture the bulk of its annual quota of rainwater, but someday it might be our only option if the climate continues to change at its current rate. Potable water is predicted to become almost as precious as gold or diamonds in many places throughout the world within just the next twenty-five to fifty years. In addition to city planners, architects are becoming increasingly aware of the need to design all types of buildings that harvest this resource.* The time has arrived to move away from the wasteful and expensive municipal habit of treating then disposing of wastewater. Some places already do recover their wastewater and reuse it (such as Orange County, California), but social marketing is crucial if the remediated water is to be consumed again. Harvesting rainwater and purifying and reusing all sources of freshwater are some of the main features that will characterize the city of the future.†

In summary, city buildings that are constructed out of engineered wood products and that harvest rainwater, generate their own energy, and grow some, or perhaps all, of the food needed to feed those living or working inside them is how I imagine the city to evolve over the next thirty years. Should all this come to pass, then cities will be comparable to intact climax hardwood forests with respect to the way they both behave: biomimicry on a grand scale.

There is nothing stopping us from fully embracing the radical changes outlined above. Success in implementing

* https://architizer.com/blog/inspiration/collections/rainwater-collection/

† E. Marx, "The future of sewage is power and profits," *Scientific American*, August 24, 2015

them does not require any new technology, only improvement to those that already exist. By becoming proactive about how the world might look and function, we can turn these ideas into global realities. Just contemplate what beautiful places the damaged portions of the Earth can once again become when we learn to live in harmony with the natural world. Many of us are not aware of the importance of how intimately connected our lives are with the rest of the nature, and I think that is why it is so commonly undervalued. For example, recent scientific evidence has revealed that the function of our microbiome (those organisms that live in and on our bodies) is to help us maintain a good standard of health. We need to do a better job of communicating this kind of knowledge to the public in order to get more minds focused on the positive side of making this a better place in which to live. We cannot afford to continue as if nothing is wrong, but if we do, then sooner or later we will suffer the same fate as the million other species whose extinction we have already caused.

Those who are not afraid to strive toward improving the human condition create the future. Helping a scarred, fragmented world to heal itself by simply leaving it alone may be the only strategy that guarantees our own long-term survival.

AFTERWORD TO THE 2020 PAPERBACK EDITION

When Dr. Dickson Despommier requested an assessment of the past ten years of the vertical-farm (VF) industry, which is one of the most interesting practices of controlled-environment agriculture (CEA) today, I chose to look forward to the exciting expectations of what is to come, as well as looking back to the actual experiences of the past. In 2000, the U.S. greenhouse industry was beginning to emerge to meet consumer demand for fresh vegetables. Several large-scale industrialized greenhouses such as Village Farms, Eurofresh Farms, and Colorado Greenhouses, beginning at ten then expanding to fifty-to-one-hundred-plus acres in size, dominated the U.S. greenhouse tomato industry. Their presence had major market impact, awakening the United States to the industry. However, the remaining producers were small family farms within three- or five- or ten-thousand-square-foot production areas, which mostly were additions to their open-field farms.

Controlled-environment agriculture (CEA), described as such in a 1967 report from the Environmental Research Lab at the University of Arizona, did not begin its modern business development to reach its current significant place in the U.S. fresh market food system and supply chain until the 1990s. During the 2000s, major inroads were made by corporate business holdings in the Northeast, Arizona, Colorado, California, and Texas, where large semiautomated monoculture CEA facilities—modeled after European farms—arose. These facilities improved tomato, cucumber, and lettuce production, enabling year-round, high-throughput harvests of name-brand vegetables.

Normally agriculture and its cultural practices evolve slowly. This was different. Continuously available produce of dependable quality became an expectation in the market. CEA greenhouses suddenly became a permanent part of U.S. vegetable production and began significantly complementing field-production agriculture, even during their high seasons.

Since 2010, a new variation of CEA has exploded onto the market because of unusual financial support from an enthusiastic and atypical business-investment community.

Instead of Silicon Valley's micromechanics technology, these investors bet on the rising towers and leafy green valleys of vertical farms.

The investment had good merit, as the market demand for safe, high-quality foods seemed sufficiently robust while the technology components existed in related forms. Although most investors were unaware, scaled systems rarely existed. Most operations were specialty designs for unique or exotic installations, not mass production. But CEA implied a controlled process, which indicated a less risky venture than traditional and unpredictable field-production agriculture. Scale-up had to begin without much experience underwriting the designs. And so there were a few stumbles, slow starts, and failed businesses.

Regardless, the amount of funds invested into VF for urban agriculture was more than could have ever been imagined, and long overdue for production agriculture in general, I would say. The plant-production environment was brought to urban areas, and urbanites wholeheartedly embraced the vegetable products within the markets. Urban agriculture can represent all types of CEA plant-production systems in all types of configurations. They are all fundamentally the same in terms of supplying the needs for growing healthy plants and plant products.

And so, the investments ensued, but not without challenges.

For the first time, new and inexperienced operations faced the everyday challenges of growing plants with liquid culture (a method in which all nutrients come from the water) within a prescribed controlled environment. They discovered that their new and improved hardware and their management technique, or "special sauce," which were beneficial to help attract investment funds, were likely not sufficient to produce saleable crops that met the market quality and volume expected from their business plan. Furthermore, the road to financial success was critically dependent on obtaining the most experienced grower possible for overseeing and directing crop production. There were few available. They were expensive. Most did not live in the United States. These operations quickly learned that the U.S. technical educational system was not able to provide educated and experienced growers. And, with such a new industry, there could be no potential growers gleaned from the sons and daughters of a previous generation.

In the late 1970s, development began on highly specialized food production for out-of-this-world applications. NASA Kennedy Space Center (KSC), along with several associated research universities, started comprehensive biological experimentation on plants, beginning with wheat, soybean, potato, lettuce, tomato, and sweet potato. The beginnings of staple and salad crops for space food. The limitations imposed by space travel and the demands of colonization on another planet led to the Biomass Production Chamber (BPC) within the NASA Controlled Ecological Life Support System Program (CELSS). The BPC provided unique experiments to support space explorers, who could not be bound to prepackaged processed food.

Engineers and plant scientists teamed together, with one focused on the needs of the plant and the other focused on the practical design limitations of space travel, to bring CEA to a new high level of technology. The BPC may be one of the earliest "indoor vertical farms," represented by a semi-automated controlled environment for plant production demonstrated in 1987 by NASA's KSC team. This confined, space-limited, highly productive system was created out of necessity, and would ultimately help feed people, recycle and reuse water, and condition the air with oxygen for the ultimate urban living. Other research facilities that followed included the European Space Agency's MELiSSA project, the EDEN ISS project in Antarctica, and the Lunar Palace in China, all currently in operation.

Just as the historically western movement of settlers throughout the North American continent included farmers, so too would the outer space movement into the solar system. This time, however, with microcomputers and sensor technology, plus energy and recycled water, these farmers would be housed within highly labor-managed and efficient spaces. The fundamental technology of the BPC was already successfully engineered. It could have supported the VF industry, but the industry was to be decades away.

While exotic, eye-catching growing systems are sometimes promoted as one prominent factor for plant growth (which of course is sold as the most important of all factors, as indicated by the promoter), there are other factors also of critical importance. It is safe to say that the multitude of plant factors must be within a reasonable range and in alignment, or one cannot improve productivity or quality of product. Certainly, one cannot immediately double yields or even increase them 10–20 percent, as many investors and their production teams have learned. Hearing such claims should be a warning to avoid the situation.

Unfortunately a plant's biological responses are unlike computer microprocessor chip performance, which, with creative design, has continually increased in capacity and speed. That was advances by physics. This is biology. The plant has numerous growth factors, and when each is increased and the plant is driven too hard, it causes reduced output and potentially self-destruction. At least with our current understanding of plant biology.

An integrated approach combining the expertise of horticulturalists and engineers can produce novel, successful systems, especially when collaborating with experienced growers. In 2016, after eleven years of experience with direct plant control, a three-member team of technologists, formerly from one of the most advanced CEA research centers in the world, Wageningen University & Research (WUR), established a practical application of a concept known as "Growing by Plant Empowerment," which carefully manages a plant's natural physiological capabilities. Through an integrated approach of physics and biology, with plant and environment monitoring and computer control, multiple plant factors are managed in precise combination. Plant "balance" is maintained for enhancing photosynthesis and transpiration by managing leaf temperature and leaf stomatal opening with net radiation, effectively keeping the plant's engine racing for optimum growth.

What's different about this approach? Having real-time information—knowledge—of the plant's condition by using sensor measurements as it responds to its environment, and then directing the computer to calculate and employ the physical environmental changes required to enhance the plant's physiological processes. An example of this is monitoring leaf temperature during daylight hours in order to control the environment and keep the leaf stomata open for CO_2 and oxygen gas exchange, thus keeping photosynthesis highly active to enhance plant growth.

If the fundamentals of both biology and physics are respected and controlled in such an integrated fashion, with a good dose of practical applied engineering, then technically the production of any crop, anywhere and at any time, is possible with CEA. There are few who have had the opportunity to experience this, and fewer who may believe this to be true, but that's changing. For food production, it will become a necessity, and many, I believe, will appreciate CEA even more in the future.

Consider the worldwide virus pandemic of 2020. Fresh food distribution systems remain dependent on geographically concentrated production of seasonal crops in open fields. Fresh vegetable production would become less tenuous if located in more uniformly distributed production sites, not only in the weather-dependent fields, but also in controlled environment facilities such as vertical farms and greenhouses, wherever they could be appropriately located.

Has it been the societal push, market demand, alternative interests, or just investors that brought urban agriculture with a focus on VF to a reality in the past ten years? Of course, it may be some combination of each, but the timing of available technology (computers, fertigation, lighting, genetics) plus the demand for safe food has made it a real option, with the economics of each specific situation yet to be fully determined.

In the past ten years, the most significant change in CEA has been the widespread use of the light-emitting diode (LED) lamp as a light source. Although the development of systems for efficient space utilization, recycling nutrient water, labor management, and automation for seeding, transplant handling, harvesting, packaging, and shipping must also be included as important factors. All have raised the technology level of CEA and VF production agriculture within a very short time.

The multiple, vertically stacked layers of plants in a vertical farm would not have been successful if not for LED lighting. The previous commercial production standard of high-intensity discharge lamps (HID), such as metal-halide or high-pressure sodium lamps, produced high operating temperatures. They emitted much heat, a limitation to plant growth when located too close. Although attempts at water-cooled HID lamps had some success, the light production efficiency of the lamp itself was not improved, whereas the LED continued to increase the conversion of electrical power to light for plant growth per input watt of electrical power.

The remaining major challenges to the success of vertical farming stem not from technical hardware, but from people. Educated employees working within a production process that is logistically sound, labor efficient, and task friendly are critical considerations for successful CEA production systems such as VF. Even with automation, the required work of skilled and educated employees will remain. Additional considerations include breeding of plant cultivars optimized for CEA production, emphasizing the importance of marketing, valuing education and experience, accepting that growing plants is a skill and an art, and understanding basic physical principles that follow the laws of physics and biology.

This is the first time in the history of the world that a new generation of food producers, that is, those actually growing the crop, can enter the industry without having to own large, or any, parcels of land. There is no need for an existing family farm. Begin in the backyard, the garage, or a spare room in the house. But knowledge and experience are required.

The University of Arizona's Controlled Environment Agriculture Center (UA-CEAC) was established with an Arizona state initiative that provided annual recurring funds beginning in 1998. The educational program became a physical reality with the first class taught in the CEAC Teaching Greenhouse, in 2000. The students enrolled in the program were expected to develop an appreciation of the interdisciplinary nature of CEA. They were provided an understanding of applied plant biology and the basic aspects of CEA engineering. This was complemented with hands-on plant production experiences with high-intensity specialty crop production in controlled environments.

Research at CEAC focused on production of edible vegetable crops. Study and demonstrations ranged from traditional greenhouses and growth chambers to applications in extreme climates such as the local Tucson summer heat, similar to many worldwide arid regions, as well as the very remote Food Growth Chamber designed for and operated by the crew of the Amundsen-Scott South Pole Station in Antarctica, and the prototype Mars-Lunar Greenhouse for a future NASA habitat on another planet.

All of these unique applications have the same common challenges to providing environmental control for effective and efficient growth based on the needs of the plant. The classroom course work and research activities helped all graduates from the UA-CEAC programs to develop an appreciation of plant production in CEA.

What would Agostino, my grandfather, who was a farmer in southern New Jersey, say about this modern agriculture? He seeded his peppers in the spring season by the date of the most recent full moon. Wood-framed hotbeds using glass sash panels as covers created the greenhouse-effect environment that warmed the seeds and the young seedlings. He knew of impending frost damage to crops in the open fields by a full moon in the fall. He also appreciated the nightfall, knowing that if farmers had light available they would work at night, too. Those memories of his successes remain significant for me. How amazed he would be to experience the modern CEA production systems, based heavily on real-time information, and, yes, electrical lighting!

I am now a grandfather. I enjoy eating fresh veggies daily, but I am removed from the traditional farmland, having only greenhouses and growth rooms at the university as my "farm." I want safe food that has consistent quality and would prefer pest control without chemicals. Everyone deserves the same; it is a human right for survival, and the basis for a stable society. Besides, I get grumpy if hungry for too long.

As I reach a career of forty years in the CEA industry, a perspective has emerged. Twenty-five years ago, the U.S. greenhouse industry was just beginning to meet some of the consumer demand for fresh vegetables by complementing the well-established U.S. field-production agriculture. Since the creation of the new name, vertical farming, there have been many food production business developments, too numerous to count. They were large and small, starting and departing, and employing various levels of technology within VF and greenhouse applications. Some have been for local food production, for meeting consumer demand for safe, fresh, high-quality food products. All have the commonality of a valuable end-product derived from the input of resources in a cost-effective manner to meet market needs and provide a profit to the business.

What will my grandchildren expect in their future? I hope that they enjoy fresh vegetables, too. With knowledge, technology, resources, food-supply chain, marketing, and financial investment, CEA will continue to help feed the world! I enjoy wine and the production of vines for grapes. I would forever be pleased that my great-granddaughter, possibly a future space traveler, would be able to enjoy a glass of cabernet sauvignon while living on the Moon or Mars. CEA technology applied to indoor growing with vertical farms can facilitate that by growing the grape to allow the craftsman winemaker to make the wine!

Gene A. Giacomelli, PhD
Professor, Biosystems Engineering Department
Former and founding director,
Controlled Environment Agriculture Center, the University of Arizona

ACKNOWLEDGMENTS

When I began to write about the vertical-farm concept, I was fortunate to be able to draw on ten years of classroom experience from a course I initiated called "Medical Ecology." In it, my students and I had already brainstormed our way through most of the serious issues created by imagining a new kind of urban agriculture and its social implications. The names of all those wonderful individuals are listed at the back of the book, and to them I am forever grateful for having had the privilege of sharing those ideas with them. I thank Lisa Chamberlain, then a freelance writer, for having the courage to interview me when we first posted the Web site www.verticalfarm.com on the internet some eight years ago, for writing an extensive, well-written piece on the vertical farm, and then, remarkably, for managing to get it published as a main feature article in *New York Magazine.* I cannot thank Steve Chen enough for being there throughout the entire process of taking the classroom concept and making it a "reality in cyberspace." He single-handedly constructed the vertical farm Web site and managed its updates with great skill and sensitivity. He remains to this day in charge of its "look." Mel Parker, my book agent, contacted me in the summer of 2008 with the prospect of producing a book about urban farming, and to him I express my heartfelt thanks for his belief in the project and continued guidance in the process of getting my ideas onto paper. His engaging daughter, Emily Parker, was enormously helpful in helping me to get this project into the hands of St. Martin's Press. I especially thank my close friend and fishing/painting companion, Robert Demarest, and his wonderfully creative wife, Alice, for their unshakable, rock-steady interest and much-welcomed constructive criticism, always offered with generous helpings of friendship and kindness, without the fear of suggesting changes where they were most obviously needed. As soon as each chapter was written, I immediately called Bob and read it aloud to him. His responses were uniformly encouraging and most helpful, though I cannot believe he still wants to fish with me, since on each trip, I predictably lead the conversation back to the contents of my book. Andrew Kranis, then a student at Columbia's School of Architecture, and I worked closely together on his final year's project, the first vertical farm design, intended as a "think piece" for the Gowanus Canal Restoration Project. His final plan was so seductive that during the charrette, the committee in charge of the fundraising effort said that they wanted his to be their first project. Thanks again, Andrew. A number of professional engineers, designers, and architects have been most helpful in my elementary education into their world of construction and design. They shared with me their insights as to the feasibility of such an undertaking as building a vertical farm from scratch: Herbert Einstein at MIT, Richard Plunz and Trish Culligan at Columbia University, Greg Kiss of Kiss and Cathcart Architects, Chris Jacobs at United Future, Eric Ellingsen at the Illinois Institute of Technology, Dan Albert of Weber Thompson Architects, and construction engineer Robert Brod. Jeanie Bochett, workplace consultant at Steelcase, New York, generously funded and arranged for a lavishly illustrated display on the vertical farm, and then installed it in their entranceway at Fifty-eighth street just off Columbus Circle, for an entire month. It attracted a lot of attention from the public and popular press alike, turning the concept from something abstract into a highly plausible idea. Thanks to the New York City Department of City Planning and the architectural firms of Grimshaw, FxFowl, and Arup, all of whom invited me to visit their inner sanctums to present the ideas that grew out of the project over the last five years. I am certain I learned much more than they did as the result of my random walks with PowerPoint. A very special thanks to all the designers, architects, and architecture students who voluntarily contributed to the visual portion of the book with highly inventive, beautifully rendered versions of their own dreams as to what a vertical farm might look like. I thank the director of The World Science Festival in New York City, Brian Greene, physicist extraordinaire, for inviting me to speak at several of its venues, greatly expanding the range of audiences that have now heard the idea for the first time. Through my participation in that enlightening series of events and presentations, I attracted the attention of speaker recruiters working for the TED (Technology, Entertainment and Design) organization, who gave me the opportunity to present the idea to yet another highly selected and appreciative audience. In that same spirit, I thank the organizers of PopTech, The Seoul Digital Forum, Taste3, and PINC for their kind invitations. I want to give a very special thanks to Stephen Colbert, who invited me to appear on his immensely popular TV show, *The Colbert Report*. He was gracious and encouraging (and very funny, too!), making sure I had time to present the idea to his viewing audience. The next day, our Web site got 400,000 hits, crashing it three separate times! I thank Exit Art and The Cooper-Hewitt museums for featuring exhibits on vertical farming, allowing yet more individuals access to the concept. Several people deserve extra-special recognition for their roles in the creation of this book. Dale Meyers, my dear art teacher and friend, read an early draft and offered excellent suggestions that are in the book today. Jake Cox arrived in my office newly graduated from Bates College and without a job. I could not offer him much in the way of a salary or a steady position, but I managed somehow to convince him that it would be worth his time in other ways if he agreed to work with me on the book. Not only did he agree to do so with joy, intelligence, and enthusiasm, but along the way he independently created a highly successful blog on the subject that is both ongoing and engaging. He (apparently, willingly) suffered through many of what must have been immensely boring days of my reading from the manuscript, again and again, as I wrote and rewrote passages, all the while, expressing his welcomed opinions on a wide variety of subjects as they arose each day. Jake proved to be good company and fun to be around. Thank you Clay Hiles for recommending him to me. Thomas Dunne Books/St. Martin's Press has been wonderful to me throughout the production and marketing phases of my book. I especially thank Tom Dunne and Sally Richardson for sharing my vision and giving it their full support. A big thanks to Marcia Markland, my editor, and Kat Brzozowski, her incredibly competent assistant, for magically transforming my words and images into a thing of beauty! Thanks also to Joe Rinaldi and Joan Higgins for their wise counseling and sage advice about all things related to the media and publicity. I thank J. D. Stettin for all his wonderful help in managing everyday things and my blog. I suspect one day he will own and operate his own vertical farm. Finally, I thank Marlene Bloom, my wife. What a fortunate person I am, for she has brought great joy to my life, and has been a solid supporter of the vertical-farm concept and this book from the moment we began to discuss its possibilities. In the summer of 1999, we "hatched" the idea and the embryo has grown ever since into a full-fledged thing of beauty. Granted, she has now become somewhat weary of the verbatim contents of my well-worn manuscript, having heard all of it many times over, but she is still the book's biggest supporter. In fact, she often offers her own version of the vertical farm and its advantages whenever the subject arises, and regardless of the venue. Her editorial skills are extensive. She has the ability to reduce a sentence to its essence and an idea to a distilled point of truth. It has served me well to listen to, and adopt most of, her constructive criticism into the body of the work that is now before you the reader. Enjoy!

APPENDICES

1) Students Who Contributed to the Vertical Farm Project
2) Author's Note on the Rainforest Fund
3) Suggested Reading
4) Web Resources
5) Additional Suggestions

Students Who Contributed to the Vertical Farm Project

CLASS OF 2004

Anisa Buck
Daniel Dine
Stacy Goldberg
Vani Gulate
Vivek Iyer
Ben Jacob
Eugene Kang
Roger Kim
Jenifer Monte
Pearl Moy
Anita O'Connor
Katerina Paraskevas
Rebbeca Tatum
Carrie Teicher
Janice Turner

CLASS OF 2005

Alam Saad
Kristen Coates
Stephen Lee
Maribeth Lovegrove
Michele Robalino
Theodore Sakata
Dennis Santella
Sapna Surendran
Kelly Urry

CLASS OF 2006

James Baumgartner
Jasmin Beria
Kenneth Chamber
Elizabeth Del Giacco
Leslie-Anne Danielle Fitzpatrick
Bryan Joshua Garber

Greg Gin
Alexis Katrell Harman
Rory E. Mauro
Jun Michael Mitsumoto
Natalie Neu
Ivan Ramirez
Elizabeth Morgan Reitano
Kathleen Ann Rosevelt
Jordana Rothchild
Nicholas Sebes
Adrienne Sheetz
Sonia Demitrie Toure
Athina Vassilakis

CLASS OF 2007

Evelyn Natalia Alvarez
Matthew Peter Bussa
Caroline Carnevale
Yana Chervona
Richele Lynn Corrado
Manisha Daswani
Jonathan Gass
Moshen Ghanefar
Katherine Gifford
Sookyung Ham
Jong Jin Jo
Dianna Jones
Steven Kauh
Raeya Khan
Danille Kontovas
Cynthia Lendor
Jason Light
Kevin Lo
Diego Lopez De Castilla
Christopher Martin
Mary Ann Popovech
Iris Anne Cruz Reyes
Yalini Senathirajah
Timon Tai

CLASS OF 2008

Sarah Autry
Claudia Cujar
Geoffrey Garst
Erica Hahn
Schuyler Henderson
Carolyn Hettrich
Yuki Kaneda
Chris Karampathis
Hannah Kellogg
Mateusz Kruk
Gilma Mantilla
Karl Minges
Christopher Ovanez
Jonathan Stettin
Sarah Wishnek

CLASS OF 2009

Joshua Bernstock
Alisa M. Koval
Yanjuska Lescaille
Sara M. Miller

Jonathan P. Salud	Alexander T. Varga
Alexander T. Sonneborn	Kate R. Weinberg
Naomi J. Sorkin	Daniel Yagoda
Sunny Uppal	Zahira Zahid

CLASS OF 2010

Patrice Adele	Lea Kiefer
Juilee Prakash Baride	Freda Robyn Laulicht
Jonathan Remy Camuzeaux	Allison Michelle Martineau
Michelle T. Chuang	Genevieve Sophia Slocum
Offira Shuly Gabbay	Ida Hui Suen
Elizabeth Ellen Hornyak	Iesha Wadala

Author's Note on the Rainforest Fund

The Rainforest Fund (www.rainforestfund.org) was established in 1989 by Sting and his wife, Trudie Styler, out of their deep passion for preserving what is left of the world's rainforests. I also am a big fan of rainforests, having visited many of them in my travels around the world. They are Earth's most beautiful places, bar none. I donate to Sting and Trudie's foundation and will continue to do so in the future. I realize that my contributions are likely a small sum in comparison to what is needed, but every little bit helps. My hope is that others who have not yet learned about their cause will also become motivated to help in the fundraising effort needed to have an impact on rainforest protection and reforestation.

Suggested Reading

CHAPTER 1. REMODELING NATURE

Defoe, Daniel. *A Journal of the Plague Years by a Citizen Who All the While Continued to Live in London.* Volume 6 of The Shakespeare Head Edition of the Novels and Selected Writings of Daniel Defoe. Oxford, England: Blackwell, 1927.

Freuedenberg, Nicholas, and Sandro Galea. *Cities and Health of the Public.* Nashville, Tenn.: Vanderbilt University Press, 2006.

Lovelock, James. *The Ages of Gala*. New York: W.W. Norton & Company, Inc., 1988.

McDonough, William, and Michael Braungart. *Cradle to Cradle*. New York: North Point Press/Farrar, Strauss & Giroux. 2002.

McHarg, Ian L., *Design with Nature*. Hoboken, New Jersey: Wiley Publishers, 1995.

www.hydroponicsfarming.com/

www.verticalfarm.com

CHAPTER 2. YESTERDAY'S AGRICULTURE

Balter, Michael. "Plant Science: Seeking Agriculture's Ancient Roots." *Science* Magazine, June 29, 2007, pp. 1830–35.

Mazoyer, Marcel, and Laurence Roudart, *A History of World Agriculture: From the Neolithic Age to the Current Crisis*. London/Stirling, Virginia: 2006.

www.archaeology.about.com/od/stoneage/ss/tishkoff_2.htm - Human migrations

www.comp-archaeology.org/AgricultureOrigins.htm

www.esciencenews.com/articles/2009/03/23/researchers.find.earliest.evidence.domesticated.maize

CHAPTER 3. TODAY'S AGRICULTURE

Brown, Stephen R. *A Most Damnable Invention: Dynamite, Nitrates, and the Making of the Modern World*. New York: Thomas Dunne Books/ St. Martin's Press, 2005.

Carson, Rachel. *Silent Spring*. New York: Houghton Mifflin Company, 1962.

Simpson, Sarah. "Nitrogen Fertilizer: Agricultural Breakthrough—An Environmental Bane." *Scientific American*, March 20, 2009.

Steinbeck, John. *The Grapes of Wrath*.

www.civilwar.com/ - civil war history

www.lightingtechnologygreenhouse.org/

www.sjgs.com/history.html - discovery of oil

www.ssbtractor.com/features/Ford_tractors.html - History of Ford tractors

www.waterforpeople.org/site/PageServer

CHAPTER 4. TOMORROW'S AGRICULTURE

Despommier, Dickson. *The Future of Our Food*. Concilience, 2010.

The State of Food and Agriculture 2009 Livestock in the balance. Food and Agriculture Organization publication. January 2010 ISBN: 978-92-5-106215-9

www.climateandfarming.org/

www.efma.org/.../Forecast%20of%20food,%20Farming%20and%20fertilizer%20Use%20in%20the%20European%20...- PDF on future of fertilizer usage in Europe

www.fao.org/worldfoodsituation/wfs-home/en/?no_cache=1

www.futurist.com/articles-archive/questions/future-of-agriculture/

CHAPTER 5. THE VERTICAL FARM: ADVANTAGES

International Symposium on High Technology for Greenhouse System Management: Greensys2007

Kenyon, Stewart and Howard M. Resh, *Hydroponics for the Home Gardner.* 6th Ed. Toronto: Key Porter Books.

Practical Hydroponics and Greenhouse Magazine (Australia)

www.aben.cornell.edu/extension/CEA/indexv4.htm

www.actahort.org/books/801/801_48.htm

www.aesop.rutgers.edu/~horteng/

www.ag.arizona.edu/CEAC/

www.hydroponicist.com/

www.thinairgrowingsystems.com/ (aeroponics)

CHAPTER 6. THE VERTICAL FARM: FORM AND FUNCTION

Gissen, David ed. *Big and Green: Towards Sustainable Architecture in the 21st Century.* Princeton Architectural Press.

Samuelson, Timothy J. *Louis Sullivan, Prophet of Modern Architecture.* New York: W.W. Norton & Company, 1998.

www.architecture.about.com/od/construction/g/ETFE.htm

www.edenproject.com/

www.lightingtechnologygreenhouse.org/ Rensselaer Polytechnic Institute Lighting Research Center

CHAPTER 7. THE VERTICAL FARM: SOCIAL BENEFITS

Christensen, Clayton M. et al. *Seeing What's Next: Using Theories of Innovation to Predict Industry Change*. Cambridge, Massachusetts: Harvard Business School Publishing, 2004.

CHAPTER 8. THE VERTICAL FARM: ALTERNATE USES

Ahmed, Iqbal, Farrukh Aqil, and Mohammad Owais, eds. *Modern Phytomedicine: Turning Medicinal Plants into Drugs*. Wiley VCH, Pubs., 2006.

Yaniv, Zohara, Uriel Bachrach, eds. *Handbook of Medicinal Plants*. Binghamton, New York: Food Products Press, 2005.

www.converanet.com/environment/living-machines-water-treatment

www.discovermagazine.com/2008/may/23-from-toilet-to-tap

http://online.wsj.com/article/SB10001424052748703503804575083611168442980.html

http://riley.nal.usda.gov/nal_display/index.php?info_center=8&tax_level=3&tax_subject=6&topic_id=1052&level3_id=6599&level4_id=0&level5_id=0&placement_default=0 At this Web site: USDA Web sites on Biofuels (PDF 88 KB).

CHAPTER 9. FOOD FAST-FORWARDED

Rainey, David L. *Sustainable Business Development: Inventing the Future Through Strategy, Innovation and Leadership*. Cambridge University Press, 2006.

CHAPTER 10. THEN WHAT HAPPENED?

Blanch, Ramiro. *Business Plan for a vertical farm Startup: Market research, business model, economic model, and more*. Beau Bassin: Lap Lambert, 2019.

Diamandis, Peter H., and Steven Kotler. *Abundance: The Future Is Better Than You Think*. New York: Free Press, 2014.

Holmes, Richard T., and Gene E. Likens. *Hubbard Brook: The Story of a Forest Ecosystem*. New Haven, CT: Yale University Press. 2016.

Kozai, Toyoki, ed. *Smart Plant Factory*. Springer Nature Singapore Pte Ltd., 2018.

Ramage, Michael H. "Supertall Timber: Functional Natural Materials for High-Rise Structures." In *Frontiers of Engineering: Reports on Leading-Edge Engineering from the 2017 Symposium*. Washington, DC: National Academies Press, 2018. Available from: https://www.ncbi.nlm.nih.gov/books/NBK481620/

Zhang, Y., D. Chen, L. Chen, and S. C. Ashbolt. "Potential for Rainwater Use in High-Rise Buildings in Australian Cities." *J. Enviro. Management* 91 (September 2009): 222–26.

www.americanforests.org/blog/forests-carbon-sinks/?gclid=Cj0KCQiApt_xBRDxARIsAAMUMu_dlFzofeZIQkKvlyQPqOsA8r-IcuevYX1zI5KxYykFxcsVi1pO5awaAsOUEALw_wcB

PHOTO INSERT FOOTNOTE

* A potential drawback of using this lightweight plastic is that it tends to leach small amounts of phthalates (potential carcinogenic compounds) when exposed to water, thereby creating a health risk (albeit a low-level one) for those who might consume food grown in this fashion. One way to prevent this is to cross-link the PVC piping prior to installing the hydroponic system. This can easily be accomplished by exposing the plastic piping to sulfides. In this way, phthalates are chemically bonded to each other and cannot leach out of the solid substrate of the plastic.

Web Resources

http://www.agreenroof.com/page8.html
Green Living™ Technologies is a privately owned company providing products and services that facilitate and simplify the integration of environmental technologies such as Green Roofs and Green Walls into our dwellings and work spaces as we empower local businesses and communities with a model that practices environmental, social, and economic responsibility.

www.agricultureinformation.com/mag
Agriculture and industry magazine

http://attra.ncat.org
Looking for the latest in sustainable agriculture and organic farming news, events, and funding opportunities? We feature all that, plus in-depth publications on production practices, alternative crop and livestock enterprises, innovative marketing, organic certification, and highlights of local, regional, USDA, and other federal sustainable agriculture activities.

www.bkfarms.com
Brooklyn Farms—hydroponic superstore

www.brightfarmsystems.com
Bright Farms Systems is a hydroponic rooftop greenhouse design and consultancy firm located in NYC.

http://cityscapefarms.com
Cityscape Farms is a hydroponic rooftop greenhouse company based out of San Francisco.

http://coolfoodscampaign.org
The Cool Foods Campaign educates the public about how food choices can affect global warming and empowers them with the resources to reduce this impact.

http://earthtrends.wri.org
EarthTrends is a comprehensive online database, maintained by the World Resources Institute, that focuses on the environmental, social, and economic trends that shape our world.

www.energysavers.gov/renewable_energy/ocean/index.cfm/mytopic=50010
Information about ocean thermal energy conversion (OTEC), a type of renewable power harnessing the battery-like method of harnessing power from big bodies of water.

http://esa.un.org/unup
World urbanization prospects

http://www.fao.org/docrep/U8480E/U8480E07.HTM
An atlas of food and agriculture

www.fas.usda.gov
FAS Mission Statement
Linking U.S. agriculture to the world to enhance export opportunities and global food security.

http://food-hub.org
Food Hub is an online network that connects local food producers to local food buyers.

www.foodinsight.org
The International Food Information Council Foundation provides food-safety, nutrition, and healthful-eating information to help you make good and safe food choices.

http://thefoodproject.org
Since 1991, The Food Project has built a national model of engaging young people in personal and social change through sustainable agriculture.

www.gardenofedenhydroponics.co.uk/home.php?cat=5
A wide variety of powered and nonpowered systems to suit every situation; kind of like amazon.com but strictly for hydroponics equipment.

www.theglobaleducationproject.org/earth/index.php
A few years ago a group of educators from British Columbia, Canada, set out to try to get an objective look at the state of the world. We wanted The Big Picture—not just this or that issue—but the most essential points of every important issue: The Executive Summary of the state of the planet. This Web site is the result of that search. The site (and the accompanying wall chart) are here to show you—in as clear, objective, and accessible a format as possible—the condition of the world, both its natural and human elements.

http://gothamgreens.com
Gotham Greens is a rooftop hydroponic greenhouse company in NYC.

www.grain.org/front
GRAIN is a small international nonprofit organization that works to support small farmers and social movements in their struggles for community-controlled and biodiversity-based food systems, mostly in Latin America, Asia, and Africa.

www.greenroofs.org
Green Roofs for Healthy Cities—North America Inc. is now a rapidly growing not-for-profit industry association working to promote the green roofs throughout North America.

www.growingedge.com
Blog on hydroponics and DIY gardening

www.hvcnyc.com
High View Creations is a rooftop garden and vertical wall garden company in NYC.

www.hydrogrown.com/reading_material_ghe.asp
Reading materials from General Hydroponics Europe

www.hydroponics.com.au/
The leading hydroponics and greenhouse magazine in the world

http://inka.fm
Inka Biospheric Systems is a socially conscious company that has created a series of solutions in response to the global "water, food, and housing" crisis.

www.justfood.org
Just Food works to increase access to fresh, healthy food in NYC and to support the local farms and urban gardens that grow it.

www.ledgrowlights.com
LEDGrowLights™ combine the gentleness of fluorescents with the growing power of HIDs. Our lights are warm to the touch and have a minimum useful lifetime of 50,000 hours—the equivalent of 18 hours a day for 7½ years.

http://www.theledlight.com
Everything made with LED lights

www.meti.go.jp/english/policy/sme_chiiki/plantfactory/index.html
Plant factories in Japan

http://www.mingodesign.com/
Green wall company

http://na.fs.fed.us/ecosystemservices/carbon/faq.shtm
Carbon market opportunities for private forest landowners

www.netafim.com/offerings/greenhouse
Netafim™ Greenhouse, one of the world's leading greenhouse solution providers, has accrued vast global experience in providing highly specialized, state-of-the-art greenhouse systems, commercial greenhouses, and greenhouse equipment.

www.nrcs.usda.gov/programs/crp/
The Conservation Reserve Program (CRP) provides technical and financial assistance to eligible farmers and ranchers to address soil, water, and related natural resource concerns on their lands in an environmentally beneficial and cost-effective manner.

www.nysawg.org
The New York Sustainable Agriculture Working Group (NYSAWG) fosters and promotes sustainable agriculture practices and sustainable local food systems.

http://our.windowfarms.org
Windowfarms are suspended, hydroponic, modular, low-energy, high-yield edible food gardens built using low-impact or recycled local materials.

www.pacinst.org
The Pacific Institute is a nonpartisan research institute that works to advance environmental protection, economic development, and social equity.

http://www.postcarbon.org
Founded in 2003, Post Carbon Institute is leading the transition to a more resilient, equitable, and sustainable world.

www.rickbayless.com/foundation/about.html
The Frontera Farmer Foundation is a nonprofit organization committed to promoting small, sustainable farms serving the Chicago area by providing them with capital development grants.

www.skyvegetables.com
Sky Vegetables is a rooftop hydroponic greenhouse company based out of San Francisco.

www.smallplanetinstitute.org/home
Living democracy, feeding hope.

www.startech.net
The mission of Startech—a plasma arc gasifier company—is to change the way the world views and employs discarded materials, what many would now call waste.

www.statemaster.com/encyclopedia/Aeroponics
Aeroponics methods overview

http://sustainableagriculture.net
The National Sustainable Agriculture Coalition (NSAC) is the leading voice for sustainable agriculture in the federal policy arena, joining together the voices of grassroots farm, food, conservation, and rural organizations from all regions of the country to advocate for federal policies and programs supporting the long-term economic, social, and environmental sustainability of agriculture, natural resources, and rural communities.

http://www.time.com/time/photogallery/0,29307,1626519_1373664,00.html
What the World Eats: a photo essay

http://www.unep.org/dewa/assessments/ecosystems/water/vitalwater/15.htm#16
Global water usage maps

www.unicef.org/sowc08/
The State of the World's Children, 2008, assesses the state of child survival and primary health care for mothers, newborns, and children today.

These issues serve as sensitive barometers of a country's development and well-being and as evidence of its priorities and values. Investing in the health of children and their mothers is a human rights imperative and one of the surest ways for a country to set its course toward a better future.

www.usda.gov/oce/commodity/wasde/index.htm
The World Agricultural Supply and Demand Estimates (WASDE) report provides USDA's comprehensive forecasts of supply and demand for major U.S. and global crops and U.S. livestock. The report gathers information from a number of statistical reports published by USDA and other government agencies, and provides a framework for additional USDA reports.

www.usda.gov/wps/portal/usdahome
USDA

www.vector-foiltec.com/cms/gb/index.php
Design and construction firm using ETFE expertly

www.worldfoodprize.org
The World Food Prize is the foremost international award recognizing—without regard to race, religion, nationality, or political beliefs—the achievements of individuals who have advanced human development by improving the quality, quantity, or availability of food in the world.

Additional Suggestions

Here are more sites to facilitate further exploration on selected topics.

http://www.aben.cornell.edu/extension/CEA/indexv4.htm

http://adsabs.harvard.edu/abs/2008AdSpR..41..730K

www.africa.ufl.edu/asq/v6/v6i3a2.htm

http://ag.arizona.edu/ceac/research/archive/hydroponics.htm

http://ag.arizona.edu/hydroponictomatoes.html

http://aquaculture-hydroponics-greenhouse.blogspot.com

www.articlesbase.com/gardening-articles/the-history-of-hydroponics-throughout-the-ages-405950.html

www.backyardfarms.com

www.baltimoreurbanag.org

www.bestgrowers.nl/start.html

www.betterbuyhydroponics.com/index.php?pr=Hydroponic_Cucumbers

www.biotech-weblog.com/50226711/hydroponics_a_smart_alternative_to_growing_rice.php

www.cahabaclub.com

www.casa-guatemala.org/map/map_location_20.html

www.commondreams.org

www.dem.ri.gov/programs/bnatres/agricult/pdf/urbanag.pdf

www.detroitagriculture.org

www.dicioccofarms.ca/index.shtml

http://edis.ifas.ufl.edu/HS147

www.eurofresh.com/default.asp

www.farmfreshri.org/learn/urbanagriculture_providence.php

www.fast3.com/index.html

www.flowgrow.co.za/companyprofile.html

www.foodsecurity.org/PrimerCFSCUAC.pdf

www.foodsecurity.org/ua_home.html

www.freshwise.org

www.freshzest.com.au

www.gipaanda.bc.ca

www.greenhousegrown.com

www.greensgrow.org

www.grow-anywhere.com

www.growhotpeppers.com/tag/hydroponic-peppers

www.growingedge.com/staff/profiles/morgan.html

www.growingpower.org

http://grumpygnome.com/journals/gardening/july6.htm

www.hos.ufl.edu/protectedag/Strawberry.htm

www.houwelings.com

http://hubpages.com/hub/Advanced-Hydroponics

http://hubpages.com/hub/hydroponicsforbeginners

www.hydroponic-guide.com/bellpeppers.php

www.hydroponic-guide.com/vegetables-carrots.php

www.hydrotaste.com

www.idrc.ca/en/ev-92997-201-1-DO_TOPIC.html

www.instructables.com/id/Hydroponic_Food_Factory/step17/Hydroponic-potatoes

www.intergrowgreenhouses.com

www.journeytoforever.org/cityfarm.html

www.k12.hi.us/~radford/vica/hydro/LETTUCE.HTM

www.kccua.org

http://lakesideproduce.com/html/overview.html

www.lans.nl/en/uk/company/mission

www.lasvegas-delight.com/index.htm

www.metroagalliance.org

www.mirabel.qc.com/default.php

www.mkeurbanag.org

www.monthlyreview.org/090119koont.php

www.mysimplehomegarden.com/garden/?p=411

www.nasa.gov/missions/science/biofarming.html

www.nbm.org/media/video/greener-good/urban-agriculture.html

www.nevadanaturals.com

www.new-ag.info/03-5/focuson.html

www.nysaes.cornell.edu/hort/faculty/weber/index.html

http://nysunworks.org

www.nytimes.com/2008/05/07/dining/07urban.html

www.nytimes.com/2009/06/17/dining/17roof.html

http://www.oardc.ohio-state.edu/hydroponics/drake/index.php?option=wrapper&Itemid=110&albumid=5035938074163502529

www.physiology.wisc.edu/ravi/okra

www.progressillinois.com/2008/09/09/.../growing-movement

www.realblueberries.com/bowerman-blueberries-hydroponic-strawberries.htm

www.redstar-trading.nl/en/uk/home

www.ruaf.org/node/512

http://savoura.com/en/section02a.html

www.seattle.gov/urbanagriculture

www.seattleurbanfarmco.com

www.sfc.ucdavis.edu/events/08workgroup/reynolds.pdf

www.sfgro.org/ua.htm

http://shenandoahrcd.org/ProjNoTillPix.htm

www.simplyhydro.com/strawberries.htm

www.smartybrand.com/index.html

www.soave.com/core/diversified_great.php

www.speedstarweb.com/other_projects.html

www.sweetwatergrowers.com

http://sweetwater-organic.com/blog

www.tcurbanag.com/

www.technologyforthepoor.com/UrbanAgriculture/Garden.htm

www.thanetearth.com/about-us.html

www.theberrypatchky.com/Berries.html

www.thinairgrowingsystems.com

www.uky.edu/Ag/Horticulture/anderson/brassica.pdf

www.urbanagcouncil.com

www.urbanagriculture-mena.org

www.urbanbliss.com/hydroponics.html

www.urbangardeninghelp.com

www.villagefarms.com

www.youtube.com/watch?v=YfVfq3lUlGM

INDEX

Index

Index

Index

Index

Index

Index

Index

Index